POWER PLAYS

ALSO BY DICK MORRIS

Bum Rap on America's Cities

Behind the Oval Office

The New Prince

Vote.com

DICK MORRIS

POWER PLAYS

WIN OR LOSE—HOW HISTORY'S GREAT POLITICAL LEADERS PLAY THE GAME

ReganBooks
An Imprint of HarperCollinsPublishers

A hardcover edition of this book was published in 2002 by ReganBooks, an imprint of HarperCollins Publishers.

HarperCollins books may be purchased for educational, business, or sales promotional use. For information please write: Special Markets Department, Harper-Collins Publishers Inc., 10 East 53rd Street, New York, NY 10022.

First paperback edition published 2003.

Designed by Kris Tobiassen

The Library of Congress has cataloged the hardcover edition as follows:

Morris, Dick.
  Power plays : win or lose—how history's great political leaders play the game / Dick Morris ; researched by Leslie Feldman.— 1st ed.
        p. cm.
  Includes index.
  ISBN 0-06-000443-6
  1. Heads of state—Case studies. 2. Political leadership—History. 3. Power (Social sciences)—History. 4. Comparative government. I. Feldman, Leslie Dale. II. Title.

JF251.M597 2002
909.82'5—dc21

                                                                2002017823

ISBN 0-06-000444-4 (pbk.)

03 04 05 06 07   RRD   10 9 8 7 6 5 4 3 2

To Terry Morris, my mother,
who taught me how to write

CONTENTS

INTRODUCTION

Politics is the pursuit of power.

History is the story of that pursuit.

First in autocracies, then in democracies, men and women have sought power through every manner of stratagem. And the history of their successes and failures offers a revealing reference for those who seek power today—whether in politics or in business, in school or the workplace, here or abroad.

This volume will examine the political moves of twenty well-known figures, probing how they sought power, why some succeeded, and why others failed. It will reach back in American history to figures as early as Abraham Lincoln and as recent as George W. Bush and Al Gore. Abroad, it will look at Tony Blair in Britain, Charles de Gaulle and François Mitterrand in France, and Junichiro Koizumi in Japan.

This book was completed as the bombs fell on Afghanistan and George W. Bush entered the most critical test of his presidency. How is he doing? To answer, I will examine how FDR and Churchill succeeded and Lyndon Johnson failed in mobilizing their nations at a time of crisis. And a special final chapter will evaluate how Bush measures up against these two yardsticks. So far, he's doing pretty well.

I've chosen the twenty men examined in this book, and their pursuit of power, because their examples have inspired, informed, chastened, animated, thrilled, and disappointed me throughout my life. There is nothing new in politics; there are only ingenious reinventions of the wheel. Any new figure looking toward a career in politics cannot help but realize that many, many others have come before, and that even a passing review of history can save more than a little heartache down the road.

I feel as though I myself have been tutored by these men—by the examples they have given us, by their struggles with strategy. Their history has taught me so much, and I write this book to share their lessons.

All politics is the same. It really doesn't matter whether one seeks the presidency of the United States, a seat as a state representative, the chairmanship of a local Rotary Club, or the presidency of one's senior class. The strategies and tactics that work for one will work for another. Some power plays take place on grand stages, with enormous audiences, but the strategic calculus they follow can be mastered by anyone interested in the pursuit of power.

Indeed, there is no politics as cutthroat as the intraoffice contest for money, prestige, and, ultimately, power. This book examines national politics, but its subtext always includes how to use the strategies that work in Washington in your own boardroom, lunchroom, or classroom.

In this survey of political tactics, I have grouped the case studies into six basic strategies:

- Stand on Principle
- Triangulate
- Divide and Conquer
- Reform Your Own Party
- Use a New Technology
- Mobilize Your Nation at a Time of Crisis

In the chapters that follow, I'll analyze successful and unsuccessful examples of each of these strategies.

Ronald Reagan, Winston Churchill, Charles de Gaulle, and Abraham Lincoln each stood on principle, endured defeat, suffered in exile, and rose again to win ultimate power in their nations. Why did they succeed? How did their heartbreaking defeats and years in the wilderness teach them to take power without having to dilute their convictions? On the other hand, why did Woodrow Wilson, Barry Goldwater, and Al Gore all fail in the end? What enables a political figure to stand on principle and succeed?

. . . And how can any individual do the same? In a world of opportunists and yes-men, how do you seize upon your own vision of the future, move toward it, and achieve the power to bring it about? What differentiates the successful man of principle and the quixotic eccentric whose path leads not to power but to self-destruction?

How did "triangulation"—that is, co-opting the opposition's issues—save the political careers of Bill Clinton, George W. Bush, and François Mitterrand? How were they able to move their parties to the center and deflate their opponents while still commanding the loyalty of their parties' orthodox wings? And failure can be just as instructive as success: When New York Republican governor Nelson Rockefeller tried the same trick by moving the GOP to the center, he succeeded only in committing political suicide. Why did he fail while the other three succeeded?

. . . And, in facing power struggles in your own sphere, how can you achieve power by stealing the other side's thunder? Can you defeat rivals by solving the problems that have given them momentum? And in making the attempt, how can you ensure that your base doesn't abandon you in the process?

Since Roman times, divide and conquer has been a proven tactic in taking and keeping power. How did it work for Abraham Lincoln, as he deliberately set out to split the Democrats over slavery, and for Richard Nixon as he stood back and let others fight over the war in Vietnam? And why did it backfire on Thomas E. Dewey, who split Truman's Democrats into three parts but still lost the election?

. . . And, too, how can one actively sow division in the ranks of one's own business rivals? What kindles division? When is ambiguity successful? And when does division not weaken but purify and fortify, as it did with Truman?

Reforming one's own political party—purging it of the elements independent voters dislike—has been a sure route to power for some like Britain's Tony Blair and Japan's Junichiro Koizumi. Why did it work for them, but blow up in the face of George McGovern in 1972 as he cleansed the Democratic Party—and our entire political system—of the pernicious influence of the political bosses?

. . . And how can self-criticism and self-reform empower one in business, or in any sphere of life? By seizing the things people criticize in your own company or group, how can you turn them around and, as a result, strengthen your supporters' loyalty? And when does this process of reformation cease to augment power, but instead amount to career suicide?

In each political generation, the first to master a new form of communication technology gains a huge advantage. FDR used radio, John Kennedy took on television, and Lyndon Johnson gave birth to the negative TV ad, each becoming the first of his era to grasp the potential of the new medium. I'll also look again at Richard Nixon, who seemed to master television in the 1950s, but fell victim to it in 1960.

. . . And how can the power of new forms of communication enable anyone to get a jump on the competition and win in any forum or contest?

Obviously, there are many and various paths to power, and many other stories from history as captivating as these twenty. But these are the strategies eight of our past twelve presidents have used to win. We need to explore them not just to learn how they did it, but to understand the lessons for us all.

Obviously, there are cruder strategies that color much of today's politics—buying the election by burying your opponent with money, for one. Finding a scandal that destroys a rival's life and career is another. Trading on a famous name is a third.

But in truth our recent history would suggest that money and scandal are *not* the real keys to real political power. After all, did Bob Dole beat Bill Clinton after outspending him two to one? Did scandal bring Clinton down?

I write in the hope that those who read this book will find its lessons helpful—and thus, as our nation mounts its momentous struggle against global terrorism, it seemed wise to add a final section to the book, offering a survey of three leaders who have faced the task of mobilizing their nations in a time of crisis, and drawing on their examples to illuminate the challenges George W. Bush confronts today. FDR's and Churchill's compelling performances at the dawn of World War II summoned the best from their peoples; Lyndon Johnson's dismal inability—or unwillingness—to explain Vietnam to the American people scuttled both the war and his presidency. As this is written, just months into the war against terror, I evaluate Bush's conduct against these yardsticks—and emerge with both assessments of his performance and suggestions of possible pitfalls to avoid as the war progresses.

My study of history is motivated by the idea that understanding the workings of power will make us all better citizens and better voters, more discerning as we watch our political leaders scramble for position and power. For only by opening the power plays of our great (and flawed) leaders to scrutiny and analysis can we gain a rich understanding of the way people work together—for their own good and for the common good.

STRATEGY ONE

STAND ON PRINCIPLE

For some leaders, the art of politics is not about movement but about positioning. More passionate about their ideas than about political gamesmanship, they hunker down and await the right moment for their ideology, deeply confident that it will certainly come. If their policies and ideas do not immediately catch fire today, so be it. If not, they are content to hang back until they sense that public opinion is on a course to converge with their vision. A position that is untenable today, such figures hope, will become inevitable tomorrow as events force things their way. The key, to them, is to monopolize and fortify their position so that they are its leading advocate and can step, uncontested, into power, proving that they had been right all along.

Like Leninists, such political leaders are confident that the future is on their side, and that they are standing on the right side of history. Ralph Waldo Emerson commented on how the worldview of such men differs from the opportunism more common in public life: "They did not yet see, and thousands of young men as hopeful now crowding to the barriers for the career, do not yet see, that if the single man plant himself indomitably on his instincts and there abide, the huge world will come round to him."

The person who chooses to "plant himself indomitably on his instincts and there abide" hopes that history will see him through his defeats to ultimate victory. As John Stuart Mill wrote, to such people "persecution is an ordeal through which truth ought to pass." Their patience in the face of adversity stems from their conviction that they are right. As Mill said: "The real advantage which truth has consists in this, that when an opinion is true, it may be extinguished once, twice, or many times, but in the course of ages there will generally be found persons to rediscover it, until some one of its reappearances falls on a time when from favourable circumstances it escapes persecution until it has made such head as to withstand all subsequent attempts to suppress it."

But history can be a faithless lover. The political graveyards are filled with men and women who have firmly planted themselves upon their "instincts," only to watch events move past them without so much as a glance over the shoulder to wave good-bye.

What distinguishes those who succeed by standing on principle from those who fail? Why does the strategy assure success for some and leave others stranded in irrelevance? What makes some who stood firm seem prescient and others appear foolish? What separates the dogmatic and stubborn who "don't get it" from the visionaries who are "ahead of their time?" It is a question that carries weight in any field of endeavor: Where does prescience end and obsession begin?

Obviously, the basic validity of the leader's vision is the most important factor. Those whose vision is fatally flawed are usually doomed to wait in vain at the political railroad station for a train that never comes. Yet even a solid grasp of the trends of history—or of the marketplace—cannot guarantee success. Many who patiently waited have gone to bitter graves, only to have their reputations posthumously honored by the vindication of their views.

There are, of course, many reasons for success and failure. But there is one factor that may distinguish those who succeed from those who do not. Those men and women who start in the wilderness, bide their time, and find that the "world comes round to them" have generally managed to weave their ideas into a deeper and more important framework, most often a vision of their nation's high calling. Those who fall short often remain entrenched in the language of ideology and fail to make the transition to the rhetoric of patriotism.

This section will examine four men who stood on principle, and ultimately succeeded: Ronald Reagan, Winston Churchill, Charles de Gaulle, and Abraham Lincoln. Each man met with initial failure and defeat, was exiled to the political wilderness, and later returned to claim power. All four kept to their principles and refused to compromise to gain office. But each, in his own way, managed the transition from ideology to patriotism as he made his final push for power.

Many felt that Reagan should have learned the futility of his hardline conservatism from the sad example of Barry Goldwater's landslide defeat in 1964. But Reagan didn't get the message. Without missing a beat, he pushed his conservative dogma with dogged consistency, through two terms as governor of California and a failed bid for the Republican presidential nomination in 1976. When Reagan ultimately returned from his four years in the wilderness and won the 1980 presidential election, little had changed in his vision—but America, in the meantime, had sunk into a national malaise under President Jimmy Carter. As Americans began to fear they'd been consigned to what Leon Trotsky called the "dustbin of history," Reagan won the presidency by wrapping his right-wing philosophy around a deep, abiding patriotism and contagious optimism about the future.

Winston Churchill's wilderness years were even longer, lasting eighteen years. Confined to the political desert in the 1920s and 1930s, this quintessential military imperialist seemed a distant voice in a world turning away from war, and a Britain turning in on itself. As the menace of Hitler grew, Churchill alone warned of the danger, decrying the policy of appeasing the Germans. But his was a lonely voice to which few listened . . . until Adolf Hitler came along to prove Churchill right, and Britons of all political stripes turned to him for leadership. Yet though his warnings were vindicated, it was not as an empire-builder or militarist that Churchill carved his enduring place in history. It was as the herald of optimism and determination in a nation almost suffocating in self-doubt and defeatism.

Looking back after World War II, Charles de Gaulle realized that party squabbling and the fractious nature of French politics had so enervated France that it contributed to its fall to Hitler in just six short weeks. Alluding to the notoriously obstinate and independent spirit of the French, de Gaulle asked: "How can you be expected to govern a country that has two hundred and forty-six kinds of cheese?" Hailed as his nation's liberator as World War II ended, de Gaulle called for a new constitution, but found himself unable to overcome the power of the political parties and resigned to await a summons to return. By 1958, the French political situation had become intolerable and the nation, at last, turned its frightened eyes to him. De Gaulle returned to power—not as the mere apostle of constitutional reform he had once been, but instead as the one man who could restore grandeur to France.

Abraham Lincoln spent a decade pleading for an end to the expansion of slavery. Too far to the left for the Illinois electorate, he won the debates but lost the election to Senator Stephen A. Douglas in 1858. After retreating in defeat, Lincoln began to shift his rhetoric, speaking more about preserving the Union and less about the moral evil of slavery. Indeed, his first presidential inaugural speech was almost wholly devoted to preserving the nation, with little mention of slavery. When he did free the slaves, more than a year after the Civil War had begun, he did so not as a moral imperative but as a military necessity in the effort to save the federal Union.

In times of stress, ideologies morph into national patriotism and they become irresistible. But the man who stands on his convictions, waiting for the world to come around, must be prepared to go patiently into the wilderness. When the consequences of his nation's, his company's, or his organization's failure to follow his advice become apparent, he must continue to imbue his message with optimism, energy, and enthusiasm. He must step forth as the hero when his time comes, not rubbing his adversary's noses in their failed ideology by saying "I told you so," but instead summoning them to join him in his crusade. Gloating, like sulking, has no place in a stand on principle.

For Woodrow Wilson and Barry Goldwater, consistent "dedication to principle" led to defeat. Each failed to wrap his ideology in patriotism and optimism and ultimately crashed.

Woodrow Wilson was an international hero after his timely intervention had saved the Allied cause in World War I. But his idealism was disfigured as he emerged bearing what became known as the worst peace treaty in history. When his vision of a League of Nations to preserve international peace met with antagonism from Congress,

Wilson took his message directly to the people, barnstorming the nation. But amazingly he insisted on defending the League in legalistic and narrow terms that failed to inspire his people. Weakened in body and spirit by the exhausting campaign, he lost sight of the compelling call of patriotism that had animated him during the war effort. The League fell along with his power and presidency.

For all their similarities of ideology, Barry Goldwater and Ronald Reagan could not have been more different in their styles. Goldwater campaigned against the liberal establishment, glorying in his contrariness. Reagan, instead, attacked America's enemies, and enshrined his program in optimism and patriotic fervor. Even the pun in Goldwater's election slogan—"in your heart, you know he's right"— seemed to stress that he spoke for only a small segment of America, not the entire nation. Where Reagan's conservatism expanded into nationalism and his ideology evolved into patriotism, Goldwater's never did. It remained pure, stubborn, clear, and unelectable.

When Gore ran in 2000, environmental issues were far down on his list. As key American voting blocs finally recognized the wisdom of his earlier warnings of ecological degradation, he lapsed into a mysterious silence on the environment, ceding the high ground— and a crucial constituency—to Ralph Nader's Green Party. The Al Gore of the 1980s would have won in 2000, but that Al Gore was long gone by the time the polls closed and the votes were counted (after a fashion).

What is it that made Reagan, Churchill, de Gaulle, and Lincoln succeed, but Wilson, Goldwater, and Gore fail in their dedication to principle? Why were the first four able to lead a successful crusade, while the last three went down to defeat?

EXAMPLE ONE—SUCCESSFUL

REAGAN STANDS ON HIS PRINCIPLES . . . AND WINS

Ronald Wilson Reagan never changed. His political philosophy was anchored in two basic principles, both linked to his belief in human freedom: the rejection of communism, and a turn away from excessive government taxation and regulation.

His entire political worldview consistently hinged on these two simple precepts. Some said it was not a sophisticated or highly intellectual approach to politics. Yet a comparison between Reagan and the only other president in the last forty years to serve two full terms—Bill Clinton—would suggest that superior intellect as a political asset comes rather overrated.

Bill Clinton's mind is a labyrinth of complexity, preoccupied with measuring ambiguities and judging shades of meaning. Every new thought is an invitation to an internalized debate in his psyche, testing its merits and attacking it from all angles. As skilled at political chess as Bobby Fischer was at the real game, he studies every possible reaction and ricochet before he makes his move.

Reagan was Bill Clinton's strategic antithesis—not least because strategy seemed almost alien to him. His mind was clear and uncluttered. The simplest of axioms governed his conduct. Where Clinton has an insatiable appetite for new notions and constant reevaluating, Reagan simply amassed ammunition to sell the ideas he already had.

And by any political accounting, Reagan was the more successful—if the less fiercely intelligent—of the two.

He held a simple view of the global war of freedom against communism. Emphatically rejecting a deal with the "evil empire," as he called the Soviet Union, Reagan wrote: "[The spectre] our well-meaning liberal friends refuse to face is that their policy of accommodation is appeasement, and appeasement does not give you a choice between peace and war, only between fight or surrender. We are told that the problem is too complex for a simple answer. They are wrong. There is no easy answer, but there is a simple answer. We must have the courage to do what we know is morally right, and this policy of accommodation asks us to accept the greatest possible immorality."

Seeing the issues of freedom and communism in clear black and white, he commented that "we are being asked to buy our safety from the threat of the Bomb by selling into permanent slavery our fellow human beings enslaved behind the Iron Curtain. To tell them to give up their hope of freedom because we are ready to make a deal with their slave masters. Alexander Hamilton warned us that a nation which can prefer disgrace to danger is prepared for a master and deserves one."

If the fight abroad was against a nation where the state ran everything, the battle at home, in Reagan's view, was very similar: to halt the government's growth and the progressive curtailment of individual liberty.

He returned to the same theme again and again.

- In 1957, Reagan inaugurated his political career with a survey of the ideological battlefield as he saw it: "This irreconcilable conflict is between those who believe in the sanctity of individual freedom and those who believe in the supremacy of the state. . . . There is something inherent in government which makes it, when it isn't controlled, continue to grow."
- As governor of California Reagan echoed the same ideas in 1967: "The time has come to run a check to see if all the services government provides were in answer to demands or were just goodies dreamed up for our supposed betterment. . . . We are going to squeeze and cut and trim until we reduce the cost of government."

- On accepting the Republican nomination for president in 1980, his message was unchanged: "It is time for our government to go on a diet."
- His farewell speech in 1988 echoed the idea that ran like a chorus through his life: "man is not free unless government is limited."

Ronald Reagan first emerged as his own man, politically speaking, after the shattering defeat of Barry Goldwater in 1964. In the wake of that debacle, Reagan was recognized as the heir to the uncertain fortunes of his party's right wing. Not that he had much competition for the role: Nobody gave the former followers of Goldwater much chance. It was assumed that his deviation from the historic norm of moderate Republicanism would prove a quixotic interlude, rather than the forerunner of successful national change that it turned out to be.

All signs in the political world had begun to shift toward the center, but Reagan would have none of it. Indeed, he told Goldwater that the Republican Party lost because it was not clear enough in articulating its position. "What we needed," Goldwater recalled him saying, "was a restatement of fundamental facts of Republicanism . . . we had lost elections because we lacked leadership, because the presence of such radical liberals as [Republican Senators] Jacob Javits, Clifford Case, Charles Mathias, and others, all wearing the Republican label, made it impossible for the voters to find any significant difference between the two major parties."

Deeply contemptuous of those who urged the Republican Party to move to the center, Reagan would later say, "I don't know about you, but I am impatient with those Republicans who, after the last election [1972] rushed into print saying 'We must broaden the base of our party' when what they meant was to fuzz up and blur even more the differences between ourselves and our opponents."

Reagan confessed that he had "always been puzzled by the inability of some political and media types to understand exactly what is meant by adherence to political principle." Rejecting triangulation, Reagan said that "a political party cannot be all things to all people. It must represent certain fundamental beliefs which must not be compromised to political expediency, or simply to swell its numbers . . . if there are those who cannot subscribe to these principles, then let them go their way."

Time and again, Reagan demanded that the Republican Party be "revitalized" with a platform that is "a banner of no pale pastels, but bold colors which make it unmistakably clear where we stand." Yet he was always eager to expand and focus his ideology, and to reach out for new allies. And he made his message work politically in a way the senator from Arizona had never done.

How did he do it? First of all, he spoke from a position of doubtlessly sincere patriotism. While Goldwater ran, unabashedly, as the head of a faction opposed to eastern liberal forces, Reagan assumed the role of the aspiring leader of an inspired nation. To Goldwater, liberals were the enemy. To Reagan, no American could ever be the enemy. The left was not evil, just deluded. Our only true enemy was the forces of tyranny.

In part, the difference was a question of timing. Coming on the heels of the national tragedy of JFK's assassination, Goldwater's squabbling seemed to take America's virtue for granted; he showed little interest in offering a positive message to a country that could have used it. Stepping forward at the end of the troubled 1970s, on the other hand, Reagan offered a welcome sense of American exceptionalism.

Reagan's conservatism was by no means purely ideological; its deepest roots were in a kind of mystical patriotism. From the very start, he reached beyond the right wing and embraced the idea of America's nationhood and destiny as his political calling. He would memorably evoke the first governor of Massachusetts, John Winthrop, and his speech "A Model of Christian Charitie," with its vision of America as "a shining city on a hill." And at a time when the country had mired for years in the tragic disappointments of Vietnam and Watergate, Reagan was the first politician since Kennedy who was able to stir in the American people a sense of their nation's mission and destiny. "Call it mysticism, if you will," he said in 1976, "but I believe God had a divine purpose in placing this land between the two great oceans to be found by those who had a special love of freedom and the courage to leave the countries of their birth. . . . We're Americans and we have a rendezvous with destiny." Ultimately, it was Reagan's ability to project his ideology as an extension of his patriotism and optimism that led America to trust him with its presidency.

Reagan's ability to invoke the essence of America, and project his faith in it, was extraordinary. Far too many leaders come to take for granted their essential mission or core beliefs. A company thinks it's trite to dwell on its commitment to service. A manufacturer assumes that everyone already understands its focus on quality. The essence of any organization—school or group or party—can become lost in a morass of details, shrouded from view by a practiced cynicism. It was Reagan's genius to rise above the dispiriting details, to wash away the cynicism, and to see America as it truly was and should be.

Reagan also added a consciousness of social morality to the largely economic basis of the Goldwater candidacy. By embracing Christian ideals of human life, he planted even deeper moral roots for his campaign.

Goldwater was never comfortable with the Christian right. His wife Peggy, a founder of the Arizona Planned Parenthood Chapter, had once declared that "Planned Parenthood is my baby." As Goldwater so delicately remarked: "I think every good Christian ought to kick Falwell right in the ass." With the pro-life movement gathering steam in the 1970s, Goldwater refused to endorse the Right to Life Amendment, although he did oppose "abortion on demand."

Reagan, on the other hand, moved to cater to social conservatives as abortion became a central national issue. Unabashedly opposed to *Roe v. Wade*, Reagan sided with the Christian right, adding a dimension to his political base that was tremendously important in growing the ranks of GOP conservatives. Seemingly contradicting his emphasis on individual choice and liberty in the economic sphere, Reagan aligned himself closely with the goals of social conservatives in his campaigns, his actions as president, and in his appointments to the United States Supreme Court.

By making the crucial alliance between these two strains of conservatism, Reagan expanded his appeal and also tapped into a growing evangelical movement in the United States. In a 1977 speech titled "The New Republican Party," he noted that he would "combine the two major segments of contemporary American conservatism . . . into one politically effective whole . . . What I envision is not simply a melding together of the two branches of American conservatism into a temporary uneasy alliance, but the creation of a new, lasting majority."

In the past, social conservatism had been more firmly allied with the left than with the right. Prohibitionists at the turn of the century, for example, worked closely with liberal advocates of woman's suffrage as wives sought the vote to ban liquor and keep their men from drinking away their paychecks. In urban areas, the Catholic Church was closely linked to the Democratic machines of New York, Boston, and Chicago in opposition to the Protestant west and south.

But as the epicenter of social conservatism shifted from the Catholic to Protestant churches, and from hierarchical movements to grassroots evangelicalism, the new Christian right found in Reagan a crucial voice welcoming them into the Republican mainstream.

By keenly grasping how his core beliefs could have new political relevance as the issues changed, Reagan acted like an established corporation launching a new product line. He traded on its traditional branding, yet tapped into a new market to meet a new need while maintaining some of its traditional formulations.

Not only did Reagan add to the Goldwater formula, he also shaved off some of its most unsavory elements as he shaped his own political coalition. Goldwater had opposed the Civil Rights Act of 1964. Seeking to ride a "white backlash" against the bill, the Arizona conservative broke into the solidly Democratic south and carried four of the states in his bid for the White House.

But Goldwater's vote marginalized him from the mainstream of his party. Even as the 1960s GOP Congressional leadership sought to court African Americans, Goldwater had pinned his hopes on support from disgruntled southern whites who were outraged at the national move toward racial integration.

While Ronald Reagan's conservatism helped him carry the South, he knew a poison pill when he saw one, and went out of his way to distance himself from racism. Where other Republicans through the years have taken positions that subtly capitalized on white fears— Nixon with his opposition to school busing and the liberal Supreme Court, George Bush, Sr., with his focus on the furlough of Willie

Horton—Ronald Reagan was largely mute on racial issues in his campaigns for the presidency.

Reagan's unwillingness to exploit racial divisions was not terribly surprising, given his longtime sensitivity to racial issues; during his political career he would recall campaigning for the integration of major league sports. Scarcely an advocate of civil rights, Reagan wasn't shy about pushing for budget cuts that fell disproportionately on the poor and minorities, earning him their everlasting animosity. But his refusal to engage in race baiting, though it did little to win him black votes, may well have made it easier for whites to back him in good conscience.

Reagan's relative open-mindedness about race may also have led him to focus increasingly on the problems of the Third World. Isolationism per se had long been exiled from the GOP, but in its commitment to fighting communist influence, the party often seemed to overlook much of the world. In the 1960s, Democrats John Kennedy and Lyndon Johnson had stressed the Third World as the battlefield of the cold war, placing an emphasis on fighting communist insurgencies in underdeveloped nations, in contrast to the Euro-focus of the Eisenhower years. After the debacle of Vietnam, however, John Kennedy's commitment to "support any friend, oppose any foe" seemed grandiose and overreaching to many on both sides of the aisle.

To Reagan, though, the specter of Vietnam was hardly daunting. As president he picked up where Kennedy and Johnson left off, focusing Republican attention on fighting communism in such distant posts as Afghanistan, Grenada, Nicaragua, El Salvador, Angola, and Ethiopia. Reagan's policy, in the words of the Heritage Foundation's Burton Yale Pines, "recognizes that national liberation movements in the Third World can be on the side of freedom and democracy and that the U.S. can help these movements win. It breaks with the 1970s when the U.S. had become isolationist and was unwilling to remain involved with the world."

By embracing social conservatives, rejecting racism, and showing concern for the Third World, Reagan updated the old conservative agenda. But it was the consistent underlying patriotism of his message, its inclusiveness and fervor, that galvanized his voters.

After Reagan rose from the ashes of Goldwater's 1964 defeat, winning the election for governor of California in 1966 and again in 1970, he seemed unstoppable in his political rise. But then he appeared to make his first big mistake—challenging President Gerald Ford for the GOP presidential nomination in 1976.

Ford, who had been appointed vice president in 1973 after Spiro Agnew's resignation and assumed the presidency after Nixon's resignation in August of 1974, was the first U.S. chief executive who had never been elected to either office. When Ford pardoned Nixon, in one of his first acts as president, most Americans suspected a deal between Nixon and Ford, and the new president's political viability quickly eroded. When Ford seemed to bumble in handling inflation, he seemed ripe for defeat.

Reagan's confidence grew as Ford faltered, and in November 1975 he jumped in to challenge the president in the GOP primaries. But Gerald Ford, tapping into his prior experience as a politician, got his act together, and his skilled chief of staff, Dick Cheney, brought a sense of order to the administration. Turned off by Nixon's paranoid secrecy and deceit, America seemed more comfortable with Ford's refreshing openness and candor. Reagan's bid faltered; after he did poorly in the early primaries, his pursuit of the nomination seemed dead.

But in the North Carolina primary Reagan found his issue, rallying supporters against a proposed treaty with Panama that would give them control of the canal that cut their nation in half. Summoning American patriotism and a commitment to putting our interests first, Reagan inveighed against the move: "We bought it, we built it, it's ours and we are going to keep it."

Reagan swept to victory in North Carolina and other GOP primaries. But his late wins were not enough to overcome the lead Ford had already built among GOP organizations. On the floor of the 1976 convention, he lost his battle by the razor-thin margin of 1,187-1,070. Undaunted, he proclaimed, "though I am wounded, I am not slain. I shall rise and fight again."

Weakened by the primaries, Ford lost to Democrat Jimmy Carter while Ronald Reagan entered what have been called his "wilderness years." Out of office, he was widely blamed for Ford's defeat. If this time was a political hiatus for Reagan, it was nothing if not instructive in the maturing of his ideology. As biographer Dinesh D'Souza notes, "Reagan's major policy agenda was based on ideas he formulated during the 70's. He recommended that the United States abandon détente and figure out a way to 'roll back' the Soviet empire" during these years. "He suggested that America return to the Monroe Doctrine and entertained schemes for getting rid of Castro. He speculated about eliminating the U.S. system of progressive taxation and returning to proportional taxation—what we now call the flat tax. He mused about the possibility of making Social Security voluntary. Some of these ideas he latter jettisoned; others he kept; all of them he tested on the American people, trusting their judgment about whether a particular proposal made sense. He was searching for the common ground between conservatism and populism."

Using periods out of power—after a fall—to recapture one's essence and shape one's message are key experiences in the careers of many who have stood on principle and succeeded. Reagan's period of reflection came just America began to feel its last reserves of optimism ebbing away.

And in this national crisis of faith, Reagan seemed to find his political raison d'être.

The sixties had been bad for America. The seventies turned out to be worse. As the decade that had begun in war and matured during Watergate ground to a dismal close, Americans began to question their own sense of national purpose, potential, and power.

Gas lines stretched for blocks; inflation reached double digits, only to be matched by high unemployment. Economists needed a new word to account for the seeming anomaly of the simultaneous miseries of an economy that wouldn't grow and inflation that didn't abate. They dubbed it *stagflation.*

When Iranian militants seized Americans as hostages and held them for more than four hundred days while President Jimmy Carter pleaded impotently for their return, the crisis seemed to epitomize how low the nation had fallen. We came to believe that the future belonged to Japan, and the theory that the United States was in decline, following the path of Britain earlier in the century, became the conventional wisdom.

Burton Pines describes the scene vividly: "It was a dreadful, terrible time in America. . . . Perhaps worst of all, we seemed to abandon the most important dream in American history: the dream that we can make tomorrow better than today . . . we were told that less is more, that resources are disappearing . . . and that yesterday was better than tomorrow will ever be."

Americans may have been dispirited, but Reagan wasn't. His optimism and courage in the face of seemingly intractable problems inspired Americans, and they flocked to his candidacy as he pledged to restore the nation's confidence.

Comparing Carter to the captain of a ship of state that "has no rudder," Reagan asked rhetorically, "Is the world safer, a safer place in which to live?" And he embodied the very sense of lighthearted optimism he was eager to mobilize among average Americans: "Recession is when your neighbor loses his job," he jibed good-naturedly. "Depression is when you lose yours. And recovery is when Jimmy Carter loses his." Indeed, the sense of optimism may ultimately have been more important to Reagan's victory than his party, his ideology, or any of his specific campaign promises.

Boldly, Reagan lashed out at the pessimism of the 1970s. No longer a mere ideologue, Reagan had turned the election into a referendum on America's potential and its future. "They say that the United States has had its day in the sun, that our nation has passed its zenith," he said in his 1980 Republican Convention speech. But he would not "stand by and watch this great country destroy itself under mediocre leadership that drifts from one crisis to the next eroding our national will and purpose."

At first Carter seemed to think he could win points by branding Reagan an extremist. Echoing the Johnson strategy that had defeated Goldwater in 1964, Carter called Reagan's brand of conservatism "a radical departure . . . from the heritage of Eisenhower and others." The strategy might have worked against an ideologue, but Reagan was coming across as a patriot and an embodiment of the American spirit. He disarmed Carter's attack in his televised debate by tiredly saying to the incumbent president: "there you go again," whenever he felt Carter was trying to misrepresent his policies, and the election tide seemed to turn in his favor.

D'Souza and others have noted the role Reagan's "instincts" played in directing his political career, and especially after Reagan assumed the presidency his essential optimism increasingly overwhelmed his strict ideology. The former actor's humor, timing, and calculated eloquence struck deep chords in our national psyche; his message of conservatism gradually morphed into patriotism, and his gallant repeated claim that "America's best days are still to come" fell on grateful ears. When the U.S. ice hockey team defeated the Soviet Union in the 1980 Lake Placid Olympics, ending the historic Russian domination of the sport, the ensuing celebration had more to do with a nation recovering its confidence than with an athletic victory.

The cycle that brought Reagan to power appears to be a constant in the lives of those who stand on principle and wait for "this huge world to come round to them." A world view rejected in one era can come to seem promising, even prescient, as the electorate changes gears. Like the beam of a lighthouse, invisible and unnecessary in the comfort of daylight, one leader's vision can emerge as a clear and welcome beacon when the fears of the nighttime come upon the land.

But the best way to understand how Reagan won in 1980 is to grasp why Barry Goldwater lost in 1964. Reagan's successes stand in contrast to Goldwater's failure, the one highlighting the other.

EXAMPLE TWO—UNSUCCESSFUL

GOLDWATER'S
CRUSADE CRASHES

Ronald Reagan was just like Barry Goldwater—and also totally different.

The two men agreed on almost everything. To read the statements of one is to hear the echo of the other. Both made opposition to communism and the Soviet Union the defining concept of their political agendas.

Anticipating Reagan's rhetoric, Goldwater made his position clear in 1964. "I am quite certain that our entire approach in the Cold War would change for the better the moment we announced that the United States does not regard Mr. Khrushchev's murderous clique as the legitimate ruler of the Russian people or of any other people. . . . Our recognition of the Soviet Union has been greatly to *its* advantage."

The men also shared a deep concern about federal spending and the growth of government. Goldwater's comments about the role of the federal government read like a blueprint for the Reagan era: "The government must begin to *withdraw* from a whole series of programs that are outside its constitutional mandate," Goldwater wrote in his famous manifesto *The Conscience of a Conservative,* ". . . from social welfare programs, education, public power, agriculture, public housing, urban renewal and all the other activities that can be better performed by lower levels of government or by private institutions or by individuals."

Both saw government as the problem, not the solution, breaking sharply from the New Deal/Fair Deal/New Frontier/Great Society consensus on the need for greater government activism. As Goldwater put it: "The need [is] for a frontal attack against . . . the Santa Claus of the free lunch, the government handout, the Santa Claus of something for nothing and something for everyone." Goldwater warned of the welfare state and its effect on individual freedom in the stark, black-and-white tones of an ideologue. "The effect of Welfarism on freedom," he wrote, "will be felt later on—after its beneficiaries have become its victims, after dependence on government has turned into bondage and it is too late to unlock the jail."

If Reagan and Goldwater agreed so much with each other, then why did the outcomes of their political efforts differ so dramatically?

Part of the answer, of course, lies in timing. The United States in 1964 was on top of its game, dominating the world economically, militarily, culturally, and ideologically. By 1980, Vietnam, Watergate, the energy and Iranian hostage crises, and the heartbreak of stagflation had taken their toll on American self-confidence. The America of 1964 was a far different place: shocked, to be sure, by the assassination of John Kennedy, yet still somehow possessed of the vitality and hopefulness he seemed to embody.

Barry Goldwater himself had a lot to do with it, too. He always seemed to have an ulterior agenda—to antagonize people. Where Reagan delighted in attracting support, Goldwater seemed to relish the role of the bad boy, spoiling the genteel lawn party with his rude comments. He said that the nation would be better off if we could just saw off the Eastern Seaboard and let it float out to sea." Revered former president Dwight D. Eisenhower, in his estimation, ran a "dime store New Deal." Republicans in Congress had failed to cut spending. Nobody got a break from the Arizona Republican.

Inside Ronald Reagan, there was joy and happiness. It came through as sunny optimism to Americans of all ideologies. But inside Barry Goldwater was a rage that expressed itself in crude volcanic outbursts, scorching friends and enemies alike.

Where Reagan grew during his years out of power, reintroduced his conservatism as Americanism, and kindled a national mood of hope, Goldwater remained a bitter ideologue, tossing bombs at the liberals and thrilling to their explosions.

But the times were not the whole answer.

Senator Barry Goldwater voted himself out of serious contention for the presidency even before launching his candidacy by opposing two of the signal achievements of the early 1960s: the 1963 Nuclear Test Ban Treaty and the 1964 Civil Rights Act. In failing to endorse these two popular measures, Goldwater isolated himself from the bulk of national public opinion. Even Ronald Reagan at his most doctrinaire never fell so completely out of touch with his country's consensus.

As senator, Goldwater voted against ratification of the Nuclear Test Ban Treaty of 1963, the U.S.-Soviet agreement to refrain from atmospheric testing of atomic and hydrogen bombs. The treaty was the first indication of a thaw in the Cold War after the near-death experience of the 1962 Cuban Missile Crisis. Widely popular, the treaty followed extensive publicity of the health dangers of the radioactive fallout generated by above-ground nuclear tests.

Flying in the face of strong public support for the treaty, Goldwater denounced it. Months later, with his campaign in full swing, the candidate compounded his problems in a televised interview on May 24, 1964, in which he wondered aloud about the potential of low-yield atomic weapons in clearing jungle supply routes in Vietnam. Working with America's first political media consultant, Tony Schwartz, the Johnson campaign capitalized on Goldwater's cavalier statements with a series of negative TV advertisements that dealt a crushing blow to Goldwater's political viability. (For more on this, see Part V.)

Goldwater also shocked America by voting against the landmark 1964 Civil Rights Act, even as he launched his presidential candidacy. The main event in a decade-long struggle against racial segregation, the act effectively guaranteed voting rights, school integration, and desegregation of public accommodations like hotels, restaurants, buses, trains, and planes. It was not the first time Goldwater had taken a stand against integration: he had also rejected the Supreme Court decision in *Brown v. Board of Education* a decade earlier, arguing that "the federal Constitution does *not* require the states to maintain racially mixed schools" and that "I am firmly convinced—not only that integrated schools are not required—but that the Constitution does not permit any interference whatsoever by the federal government in the field of education. It may be just or wise or expedient for negro children to attend the same schools as white children, but they do not have a civil right to do so which is protected by the federal constitution, or which is enforceable by the federal government."

As if obstinately ignoring the overwhelming national consensus about nuclear arms and peace were not enough, Goldwater was also the first modern candidate to risk touching the "third rail" of American politics—Social Security—by recommending that participation in the system be voluntary. "If a person can provide better for himself . . . let him do it. But if he prefers the government to do it, let him." Almost immediately the press attacked him. The Concord *Daily Monitor,* for example, greeted New Hampshire voters with a banner headline: "Goldwater Sets Goals: End Social Security." The Johnson campaign jumped on the issue and produced an ad that showed a Social Security card being torn up into little pieces, stoking elderly fear over the Republican's plans for their lifeline.

Armed with its rival's three faux pas—civil rights, nuclear testing, and Social Security—the Johnson campaign was able to paint a continuing picture of Barry Goldwater as a hotheaded extremist. As Goldwater himself later complained, "My opponents built a caricature of Goldwater, and this was used by both my Republican primary opponents and Mr. Johnson in the general election campaign. This caricature was built upon a 'trigger-happy' fellow and 'the man who would tear up Social Security cards.' Both of these premises were completely false. But this thing started in the primaries—and, try as I would, it could not be erased. . . . It was entirely a fear campaign."

Was Goldwater an unlucky ideologue, or the victim of his own mule-headedness? Was his refusal to account for public opinion a matter of conviction, or willful disregard of political realities?

Again, the comparison with Ronald Reagan is instructive. Reagan had one virtue Goldwater could never manage: he appreciated the advantages of silence when speaking out could be hazardous to one's political health. After House Democratic Speaker Tip O'Neill defeated his plan to trim cost-of-living adjustments, for example, Reagan refused to raise the question of toying with social security. He recognized the difference between consistency and suicide—a distinction that was lost on Senator Goldwater.

If there was a single reason that Goldwater fell so hard in 1964, it was that his campaigning style lacked that unifying, positive tone Reagan brought to his conservative crusade. Compare Goldwater's and Reagan's speeches, and the negativism and factionalism of the former stands in sharp contrast to the high-spirited patriotism of his successor. Nobody escaped the expostulations of what the *New York Times* politely called Goldwater's "sometimes intemperate personality." Goldwater couldn't even restrain himself from heaping scorn on a host of GOP icons during his campaign. He blamed Henry Cabot Lodge and Nelson Rockefeller for the Republican Party's defeat in the 1960 presidential election, saying that "If either—and preferably both—had worked half as hard as the rest of us did, Mr. Nixon would be in the White House today."

He savaged Eisenhower's longtime chief of staff, saying that if elected his first task would be "to restore the strength of the Republican National Committee to what it was before Sherman Adams destroyed it."

Goldwater criticized fellow Republicans Robert Taft and Eisenhower, accusing them of failing to keep their promise about reducing government spending. "Now it would be bad enough if we had simply failed to redeem our promise to reduce spending; the fact, however, is that federal spending has greatly *increased* during the Republican years." And of course he was unstinting in his critique of Lyndon Johnson, calling him "the biggest faker in the United States" and the "phoniest individual who ever came around."

Some of his fellow Republicans were horrified by his rhetoric. Nelson Rockefeller called his statements "dangerous" and "frightening." New York City mayor John Lindsay told the *New York Times* that he was "searching his conscience" before deciding whether he could support Goldwater. Even former president Eisenhower was appalled, saying that Goldwater's speech was "giving the right-wing kooks a pat on the back and everyone else a slap in the face."

A simple comparison of Goldwater's convention speech with that of Ronald Reagan throws Goldwater's bear-baiting tendencies into high relief. Goldwater's speech was focused on party and ideology; he mentioned the word *Republican* thirty-two times in his 3,100-word address, whereas Reagan used it only four times in 4,900 words (including once while quoting Lincoln and once in an appeal to "Republicans, Democrats, and Independents").

Goldwater dwelled on ideological themes for 37 percent of his speech, while Reagan devoted just 22 percent of his remarks to such topics; nearly half of Reagan's speech was devoted to evoking national pride, while Goldwater dealt with such emotions fleetingly at best.

Reagan extended an olive branch toward the Democrats, quoting from FDR himself, and described the GOP as "a party ready to build a new consensus with all those across the land who share a community of values embodied in these words: family, work, neighborhood, peace, and freedom." His goal, he said simply, was "to unify the country, to renew the American spirit and sense of purpose."

Goldwater's address, on the other hand, was a purely partisan, conservative manifesto. He condemned a government philosophy that would "elevate the state and downgrade the citizen." He said that equality "rightly understood . . . leads to liberty," but that "wrongly understood, as it has been so tragically in our time, it leads first to conformity and then to despotism." He pledged to "resist concentrations of power, private or public, which enforce such conformity or inflict such despotism."

His speech featured a paean to private property, and called on government to dedicate itself to fostering it. "We see in the sanctity of private property," he said, "the only durable foundation for constitutional government in a free society."

Linking liberty to "decentralized power," Goldwater went on to condemn "those who seek to live your lives for you" as the precursors of those who will substitute "earthly power" for "divine will." Attacking Johnson as lustily as Reagan went after Carter, he was not content with merely criticizing his opponent's performance. He insisted the difference was more profound. "Now, we Republicans see all this as more, much more, than the rest: of mere political differences or mere political mistakes. We see this as the result of a fundamentally and absolutely wrong view of man, his nature and his destiny."

Where Reagan welcomed all into his national crusade, Goldwater was more circumspect. "Anyone who joins us in all sincerity, we welcome. Those who do not care for our cause, we don't expect to enter our ranks in any case. And let our Republicanism so focused and so dedicated not be made fuzzy and futile by unthinking and stupid labels."

Finally, one can only gasp at the self-destructiveness of Goldwater's unforgettable declaration at his convention that "extremism in the defense of liberty is no vice. And let me remind you also that moderation in the pursuit of justice is no virtue." In that one gesture, Barry Goldwater had effectively marginalized his candidacy nearly into extinction.

As historian Michael Gerson notes, "Reagan and Goldwater both represented the Western spirit, but Goldwater was the maverick and outsider to Reagan's easygoing hero. One had a talent for honesty, the other, it's been said, a talent for happiness. It is obvious what America rewards. Goldwater captured the party. Reagan captured the country, largely by blunting the sharp angles of '64."

Goldwater paid dearly for his apostasy in attacking his fellow Republicans, none of whom appreciated his wit or sarcasm. Pennsylvania governor William Scranton, in his late challenge to Goldwater's nomination, called the Arizonan's philosophy a "crazy quilt collection of absurd and dangerous positions" and warned that Goldwater "too casually prescribed nuclear war as a solution to a troubled war." When Nelson Rockefeller and others sought to undo the damage by calling for a formal disavowal of extremism, the resulting furor among delegates on the floor created a televised spectacle that frightened America and summoned eerie memories of the fascist brownshirt tactics of decades earlier. (Defeated at the Republican convention, the plan was eventually adopted by the Democrats.)

Future president George H. W. Bush described the peculiar character of the supporters that made up Goldwater's constituency in a postmortem on the latter's candidacy that ran in the *National Review*. Bush blamed Goldwater's "negative image" on the "so-called 'nut' fringe." As Bush recalled, "an undecided voter would be pounced on by some hyper-tensioned type armed with an anti-LBJ book or an inflammatory pamphlet. The undecided voter wouldn't get a sensible message on where Goldwater stood, he'd get some fanatic on his back tearing down Lyndon. Goldwater didn't want to repeal Social Security but some of his more militant backers did. He didn't want to bomb the UN but these same backers did. They pushed their philosophy in Goldwater's name, and scared the hell out of the plain average non issue conscious man on the street."

At the time, Goldwater's defeat seemed a result of his extreme views—and, as Bush notes, those of his supporters. Obviously these issues played a large part in the debacle. Had Goldwater voted for the Nuclear Test Ban Treaty and the Civil Rights Bill and not mused publicly about voluntary social security and tactical nuclear weapons, he would have been far more electable.

Forced onto the defensive by his own choices and remarks, Goldwater spent the entire campaign defending himself against the charge of extremism. According to Goldwater, "The whole campaign against me was run on fear of me." It was, by any measure, his own fault: Where Reagan's charming way of mocking his opponents lifted the tenor of the 1980 election season, Goldwater's campaign was weighed down by the candidate's abrasive style, his penchant for making enemies, and his failure to evoke any meaningful sense of unified mission among average Americans. It was a recipe for rejection: Goldwater lost and lost badly, getting only twenty-seven million votes to Johnson's forty-three million.

What are the lessons of Goldwater's candidacy? One is certainly that style matters. The negative, combative tone Goldwater brought to his campaign—the sense that there were enemies around every corner—did much to undermine his appeal. This is especially true for the man (or woman) of principle who would insist on a major change of course: The successful ideologue must be more ingratiating than the opportunist and more charming than the charlatan. Only by marshaling every political skill can he turn consistency into victory.

Rather than transcend his own ideology and embrace the credo of national pride and persistent optimism as Reagan did, Goldwater defined his campaign narrowly, within the four walls of his ideology—and ultimately expired within its bounds.

CHURCHILL EMERGES FROM THE WILDERNESS TO LEAD BRITAIN IN ITS FINEST HOUR

Winston S. Churchill had to wait more than thirty years for the world to come around to him. Like Reagan, he never budged from the positions he had embraced while in the wilderness. It was the world that came to him. But it needed a little prompting.

Most biographies devote relatively little time to Churchill's deliberate campaign to take power once World War II began. For eight months after war was declared, the world mocked it as "phony" because the major armies had not yet become engaged. But Churchill, as First Lord of the Admiralty, was busy campaigning for the post of prime minister. As surely as any politician stumping for office, Churchill galvanized his support among the British people through a series of stirring radio addresses that contrasted sharply with the lackluster image of the worn-out incumbent, Neville Chamberlain.

Bringing his optimism, vigor, determination, wit, and élan to the attention of the average person on the street, Churchill finally attracted the popular support that had eluded him for his entire career.

Without radio, Churchill might never have become prime minister. It was the ability to bring his voice into British homes—charming, enlightening, and inspiring all who heard him—that helped him reach the summit. And it was radio that allowed him to sustain

Britain's morale and win the war. President John F. Kennedy said of Churchill that he "mobilized the English language and sent it into battle." Churchill's story illustrates the crucial role persona can play in politics. Positions alone do not a leader make. Nowadays, when personality politics is much maligned, it is well to look back to the dark days of World War II and realize how crucial character, charisma, and personality can be in fighting the good fight.

For decades before the war, Winston Churchill seemed to most of the British people to be a man out of place—an anachronism from an era of empire, an imperialist before the moniker became an accusation.

Churchill had first burst on the British political scene in 1901, when he filed gripping news dispatches boasting of his heroic escape from enemy clutches during the Boer War in South Africa. A young man in a hurry, he appeared to be on his way to the top.

When World War I began, he was appointed to the prestigious position of First Lord of the Admiralty, the traditional seat of British military power. Horrified by the mindless slaughter during the war in France, where men died by the hundreds of thousands for a few yards of territory, he urged a flanking movement around the stalemated battlefields by attacking southern Europe through the Dardanelles.

Churchill got his expedition—but it was a disaster, for which he was only partly to blame. After sustaining almost three hundred thousand British and Allied casualties, the mission was forced to abandon any hope of occupying the Dardanelles.

Churchill took all the responsibility on his own head. "I have had the blame for everything that has gone wrong," he said. "I have done my best." British politicians were quick to demand his head. When he was demoted in disgrace, Churchill's wife feared "he would die of grief."

Churchill survived, but his political career seemed lost forever. When appendicitis prevented him from appearing in public during the 1922 campaign, he lost his seat in Parliament. He moaned that he was "without an office, without a seat, without a party, and without an appendix."

His wilderness years had begun. He seemed a man of war in a time of yearning for peace.

In the 1920s, it appeared as if the world had finally learned its lesson about the futility and barbarism of war. As virtually every nation signed the idealistic Kellogg-Briand Pact, which sought to outlaw war, and naval powers agreed to pare down their fleets to fixed ratios, peace was on the march.

Churchill was badly out of step. J. R. Clynes, in his 1937 *Memoirs*, recalls the fallen leader's militarism: "Churchill was, and has always remained, a soldier in mufti. He possesses inborn militaristic qualities, and is intensely proud of his descent from [the Duke of] Marlborough. He cannot visualize Britain without an Empire, or the Empire without wars of acquisition and defence." A mere two years before the start of World War II, Clynes wrote that "a hundred years ago [Churchill] might profoundly have affected the shaping of our country's history. Now, the impulses of peace and internationalism, and the education and equality of the working classes, leave him unmoved."

Another Churchill associate, Kingsley Martin, also found Churchill's views out of sync with the times: "He was a delicious and witty guest, quite willing to talk freely to young academics. I then regarded him as the most dangerous of all politicians. He combined brilliance with the most foolish and antiquated views, which would have condemned us without hope of reprieve to war between classes and nations. . . . All the more remarkable that I was to become his admirer in the later thirties and to write a eulogy of him as our indispensable leader in 1940."

Churchill's imperialism was especially fervent when it came to India, which he desperately wanted to keep in the British Empire. His attitudes revealed a weakness in his character—a condescension to the peoples of the subcontinent and the Far East: "In dealing with Oriental races," he said, "it is a mistake to try to gloss over grave differences, to try to dress up proposals in an unwarrantably favourable guise, to ignore or conceal or put in the background rugged but unpleasant facts. The right course on the contrary is to state soberly and firmly what the British position is, and not be afraid to say 'this would not suit us,' 'that would not be good for you,' 'there is no chance of this coming to pass,' 'we shall not agree to that being done.'"

But whatever its excesses, ultimately it was his vision of the great British Empire that was Churchill's defining passion. He spoke of proclaiming "to all the world that the heart of the Empire is true and that its hand is just and strong." His preoccupation with Empire and national grandeur diminished his popularity as the appetite for peace gripped the nation in the tragic aftermath of World War I, and he was consigned to speaking for a faction within a faction of British politics. But, like Ronald Reagan, Winston Churchill never wavered from his personal convictions to fit the mood of the moment.

Returned to Parliament, Churchill watched the growth of Nazism in Germany, fascism in Italy, and militarism in Japan with growing apprehension. Supported by only a handful of followers, he spoke urgently of the danger of increasing German military strength, and especially of the threat the burgeoning Luftwaffe posed to the security of the British Empire.

But the British weren't listening, at least not at first. Churchill's cries seemed shrill to a nation with its head in the sand. Unable to face the horror of another world war, the British public tuned out his dire forecasts of Nazi strength. Prime Minister Stanley Baldwin dismissed demands for naval armament, calling England's navy an "expensive toy." When Churchill warned about the growing imbalance between German and British arms and its possibly horrific consequences, Neville Chamberlain, then the Chancellor of the Exchequer, "blocked everything on the grounds of finance," imperiling Britain's defenses to balance her budget.

Throughout England, disarmament became the national passion. Oblivious to the threat posed by Hitler, politicians and the public demanded a reduction, not an increase, in arms production. Churchill mocked their emphasis on arms control by concocting a children's bedtime fable in which bears and rhinos and lions and zoo animals voluntarily gave up their teeth and horns and claws, but not their mutual animosity. Once disarmed, the tale reveals, the animals nevertheless fought with whatever natural equipment they retained.

With disarmament, inevitably, came appeasement—the widely held conviction in the Britain of the thirties that Hitler would be placated if only he were given Austria and Czechoslovakia. Churchill heaped sarcasm on the advocates of appeasement, who "think . . . the English hopeless and doomed" and who proposed "to meet this situation by grovelling to Germany. 'Dear Germany, do destroy us at last!' " Churchill fumed. "I endeavor to inculcate a more robust attitude." As British complacency grew around him, his warnings only grew starker. Addressing the House of Commons in May 1935, he tried to awaken his colleagues: "One would imagine, sitting in this House today, that the dangers were in process of abating. I believe that the exact contrary is the truth—that they are steadily advancing upon us," he forecast grimly. "Nourish your hopes, but do not overlook realities."

To an outside observer, Churchill would have seemed shut out of political power just as surely in the late 1930s as he had been throughout the previous fifteen years. Those in power still loathed him, and the British public continued to turn a deaf ear to his pleas.

But as the decade unfolded Churchill began to realize that he had a strong ally on his side—the truth. The world, courtesy of the threat Adolf Hitler posed, was beginning to come around to him. Rearmament, which seemed mere folly and paranoia in the 1920s, began to appear prudent, indeed overdue.

As Churchill watched Chamberlain lead Britain into weakness and military inferiority, he must have felt much as Reagan did forty-five years later, watching Carter emasculate America's strength and defenses. For Churchill—as for Reagan—the anguish they felt must have been tempered by faith that their countrymen would soon see the light and heed their voices.

The final stand for appeasement, of course, came when Prime Minister Neville Chamberlain journeyed to Munich to negotiate with Hitler in an effort to avert war. Chamberlain, a pacifist, was awed by the massive display of strength the German dictator staged for his benefit. After caving in to Hitler's demands for control of a large part of Czechoslovakia, Chamberlain returned, umbrella in hand, promising that the accords he had negotiated would assure "peace in our time."

London erupted in a frenzy of relief. Churchill biographer Martin Gilbert described the spirited welcome that greeted the prime minister: "In the five days since Chamberlain's return from Munich there had been much public rejoicing. Almost every newspaper had been ecstatic in his praise."

But Churchill was not fooled. Condemning Munich as a "total and unmitigated defeat" and "a disaster of the first magnitude," he pressed for a national unity government prepared to confront the Nazi menace. "The partition of Czechoslovakia under pressure from England and France," he contended, "amounts to the complete surrender of the Western Democracies to the Nazi threat of force . . . [It] will bring peace or security neither to England nor to France. On the contrary, it will place these two nations in an ever weaker and more dangerous situation."

When, predictably, the Germans proceeded to occupy the rest of Czechoslovakia, violating their pledge at Munich, it was clear that war was coming. At last Britain and France drew a line in the sand, proclaiming that an attack on Poland would lead to world war.

And yet to many, Churchill still seemed too dangerous to hold a top position in the government. His bellicosity frightened Britain, and worried its diplomats. Chamberlain continued to govern without regard for Churchill, although the latter knew his time would come. "I understood the Prime Minister's outlook," Churchill remembered later. "He knew, if there was war, he would have to come to me, and he believed rightly that I would answer the call. On the other hand, he feared that Hitler would regard my entry into the Government as a hostile manifestation, and that it would thus wipe out all remaining chances of peace. This was a natural, but a wrong, view."

As war loomed, the demand for Churchill increased. As he happily remembered, "thousands of enormous posters were displayed for weeks on end . . . [saying] 'Churchill Must Come Back.' Scores of young volunteer men and women carried sandwich-board placards with similar slogans up and down before the House of Commons."

When the other shoe dropped, and Germany invaded Poland, Britain and France had no choice but to declare war. With that prospect before him, Chamberlain overcame his reluctance and invited Churchill into the government as First Lord of the Admiralty—the very same position he had held during World War I.

As First Lord, Churchill was powerful, but he was not Prime Minister. Without control of the government, he knew he would not be able to act effectively. And so now, Churchill began an eight-month political campaign—though he would never admit it—to achieve power.

As word spread throughout the Admiralty that "Winston is back,"

First Lord Churchill sent out so many memos with the heading "pray inform me" or "pray send me" that they became famous as the "First Lord's Prayers."

Chamberlain, of course, was deeply unprepared for the challenges facing England. His age, lethargy, and commitment to peace at any cost made it impossible for him to rally the nation. An observer recounts the attitude of the moment: "The Prime Minister gets up to make his statement. He is dressed in deep mourning. . . . One feels the confidence and spirits of the House dropping inch by inch. When he sits down, there is scarcely any applause. During the whole speech, Winston Churchill had sat hunched beside him looking like the Chinese god of plenty suffering from acute indigestion."

When Churchill's turn came to speak, however, the reaction was very different: "The effect of Winston's speech was infinitely greater than could be derived from any reading of the text. . . . One could feel the spirits of the House rising with every word. . . . In those twenty minutes, Churchill brought himself nearer the post of Prime Minister than he has ever been before. In the Lobbies afterwards even Chamberlainites were saying, 'We have now found our leader.'"

A public opinion poll taken in December 1940, four months after the war began, found barely half of the British people clinging to confidence in Chamberlain. In *The Last Lion,* his splendid account of Churchill's long years out of power, William Manchester quotes one disillusioned Conservative as saying that Chamberlain was "hanging onto office like a dirty old piece of chewing gum on the leg of a chair."

Except for Churchill, the British government was rife with the appeasers whose policies had failed to prevent war. Typical of their attitude was the response of the air minister to a suggestion that Britain drop "incendiary bombs on the Black Forest" to support Poland as it battled Nazi aggression in the East. "Oh, you can't do that," the air minister said, "that's private property. You'll be asking me to bomb the Ruhr [the center of German industry] next." Pressed to aid the Polish resistance with raids on German industry, Chamberlain chose instead to drop leaflets over German cities pointing out Hitler's atrocities in the hope of inducing surrender, a mission that led to the loss of British aircraft and fell predictably short of its objective.

As Manchester notes, "Before England could fight, she needed . . . a government of fighting ministers. . . . Churchill was such a man. Despite his membership in the cabinet, however, he was virtually alone. The rest of the government was schizoid. Their faith had failed . . . [but] the appeasers were still devout, still hopeful that the shopworn messiah at No. 10 [Chamberlain] would be vindicated."

As the British clung to their policy of appeasement even after the war had begun, Churchill was indignant. He later described his feelings at the time: "This idea of not irritating the enemy did not commend itself to me. . . . Good, decent civilised people, it appeared, must never strike themselves till after they have been struck dead. In these days, the fearful German volcano and all its subterranean fires drew near to their explosion point. . . . On the one side endless [British] discussions about trivial points, no decisions taken, or if taken, rescinded, and the rule 'don't be unkind to the enemy, you will only make him angry.' On the other [German side] doom preparing—a vast machine grinding forward ready to break upon us."

Determined to assume the prime ministership, Churchill embarked on his full-scale, if undeclared, campaign to win the top job. Radio was his weapon. As Manchester describes it, Churchill used the medium with passion: "Until he entered the War Cabinet, Churchill's audiences had been largely confined to the House [of Commons], lecture halls, and, during elections, party rallies. Suddenly that had changed. England was at war; the only action was at sea, and millions whose knowledge of Churchillian speeches had been confined to published versions heard his rich voice, resonant with urgency, dramatically heightened by his tempo, pauses, and crashing consonants, which, one listener wrote, actually made his radio vibrate. Churchill had been a name in the newspapers, but even his own columns lacked the power of his delivery. He found precisely the right words for convictions his audiences shared but had been unable to express."

They weren't campaign ads, not even stump speeches. But Churchill's BBC addresses showed Britain that there was an alternative to the faltering Chamberlain. For Churchill, it was his opening battle in the war: he knew it fell to him to defend Britain, but he knew first that he must win the hearts and minds of the British people.

Radio was the medium that allowed Churchill to deploy his magnificent personal style in service of his message. Like so many other leaders in times of crisis, he knew instinctively that the ability to communicate with people on an emotional level was essential to his political success. Speeches are not meant to be read, they must be heard. When Winston Churchill began to speak directly to the British over the airwaves—as only a cabinet member had the privilege to do—he was no longer the troglodyte imperialist of old, but a warm and charming patriot with a manner at once ingratiating and inspiring.

Churchill's addresses were filled with drama and triumph. In one October 1939 BBC address, he held the nation on the edge of its seats as he described his Navy's encounter with German submarines. The U-boats "sprang out upon us as we were going about our ordinary business," he informed listeners, "with two thousand ships in constant movement . . . upon the seas." Though "they managed to do some serious damage," he added, his voice strong and confident, "the Royal Navy . . . immediately attacked the U-boats and is hunting them night and day—I will not say without mercy, for God forbid we should ever part company with that—but at any rate with zeal and not altogether without relish."

When Hitler had finished his conquest of Poland, he extended an offer to negotiate peace with the Western democracies. Churchill took to the radio to denounce it on November 12, 1939: "We tried again and again to prevent this war, and for the sake of peace we put up with a lot of things happening which ought not to have happened. But now we are at war, and we are going to make war, and persevere in making war, until the other side have had enough of it. . . . You may take it absolutely for certain that either all that Britain and France stand for in the modern world will go down, or that Hitler, the Nazi regime, and the recurring German or Prussian menace to Europe will be broken or destroyed. That is the way the matter lies and everybody had better make up his mind to that solid, somber fact."

No longer was Churchill preoccupied with the virtues of empire-building. Now he was battling for the very survival of Britain itself. He concluded one BBC broadcast with a ringing call to arms: "Now we have begun; now we are going on; now with the help of God, and the conviction that we are the defenders of civilisation and freedom, we are going on, and we are going on to the end." The British people had a leader once more.

Straying afield of his nominal role as First Lord of the Admiralty, Churchill told the nation, via radio, that Germany was vulnerable. He spoke of having "the sensation and also the conviction that that evil man over there and his cluster of confederates are not sure of themselves, as we are sure of ourselves; that they are harassed in their guilty souls by the thought and by the fear of an ever-approaching retribution for their crimes, and for the orgy of destruction in which they have plunged us all. As they look out tonight from the blatant, panoplied, clattering Nazi Germany, they cannot find one single friendly eye in the whole circumference of the globe. Not one!"

Churchill was speaking to rally his nation, but he was also mounting a passionate campaign to be elected prime minister—and, as Manchester describes, he was making headway. "Among the middle and lower classes, pacifism had begun to fade when Hitler entered Prague, and once war was declared it was replaced by patriotism. . . . the signs of the nation's shift in mood had been unmistakable. . . . There was talk—more out of Parliament than in it—of Churchill as prime minister."

Churchill's warlike attitude was sounding better and better to Britain. As William Manchester wisely observed, "His natural aggression, curbed in peacetime, a stigma only a year earlier . . . [was] now a virtue." As Manchester noted, even Chamberlain's junior private secretary, Jack Colville, confided to his diary that the First Lord of the Admiralty was making progress in his campaign to become prime minister. Churchill, Colville wrote, "certainly gives one confidence and will, I suspect, be Prime Minister before this war is over. . . . he is the only man in the country who commands anything like universal respect."

As it happened, even Adolf Hitler agreed. Before long he began to lash out at Churchill by name, even though his antagonist was still just head of the Navy. When the Führer deigned to offer Britain and France peace after he swallowed Poland, he told the British public that it had to choose between peace and "the views of Churchill and his following." Hans Fritzsche, director of the Nazi broadcasting services, was more explicit in his denunciations: "So that is what the dirty gangster [Churchill] thinks! Who does that filthy liar think he is fooling? . . . So Mr. Churchill—that bloated swine—spouts through his dirty teeth that in the last week no English ship has been molested by German submarines? He does, indeed?"

When there was good news to offer, such as the sinking of the German battleship *Graf Spee,* Churchill hastened to the radio to enthrall Britain with his graphic account. The German ship boasted "eleven-inch guns" with a range of "fifteen miles." The terror of the seas, it "sent nine British cargo ships to the bottom" since the war had started only one hundred days earlier. On December 13, 1939, the ship was discovered off South America by a trio of Allied ships, the *Exeter,* the *Ajax,* and New Zealand's *Achilles.* Although the Allied trio were badly outgunned and suffered great damage, they crippled the German vessel, wounding its captain, killing thirty-seven of its crew, and opening "gaping holes" in its structure with eighteen hits. The *Graf Spee* "limped" into a neutral harbor in Uruguay, where its captain scuttled it and killed himself.

Churchill later wrote that the destruction of the *Graf Spee* "gave intense joy to the British nation and enhanced our prestige throughout the world. The spectacle of the three smaller British ships unhesitatingly attacking and putting to flight their far more heavily gunned and armoured antagonist was everywhere admired."

Chamberlain, in the meantime, could do nothing right. Churchill proposed, early in the war, that Britain occupy and mine the harbor of Narvik in Norway to block exports of Swedish ore from reaching Germany. But from the beginning, the operation was bungled. The French were slow to approve of the mining operation and the troops Britain sent were inexperienced Territorial Forces, since their best men were busy shoring up the defense of France. The sea lords around Churchill—and the man himself—underestimated the impact of air power on naval operations. But, most important, Hitler got the same idea at the same time, and sent massive German forces to occupy Norway before the Allies could get there.

Yet in the end there was no time to parse out blame for the disaster, for it was soon subsumed in the far larger catastrophe that followed swiftly: the invasion of France.

As the German army descended on France from the north, cutting off British and French troops in Belgium and the northern provinces, it became clear that Britain was truly facing its darkest hour. With the war raging, there was no time or need for an election. Now, at long last, Chamberlain stepped aside and handed the reins to Churchill—who formed a unity government with Labour and Tory ministers that would last through the entire war.

As Churchill later described his emotions on taking power: "on the night of the tenth of May, at the outset of this mighty battle, I acquired the chief power in the State. . . . During these last crowded days of the political crisis, my pulse had not quickened at any moment. I took it all as it came. But I cannot conceal from the reader of this truthful account that as I went to bed at about 3 A.M., I was conscious of a pro-found sense of relief. At last I had the authority to give directions over the whole scene. I felt as if I were walking with Destiny, and that all my past life had been but a preparation for this hour and for this trial. Eleven years in the political wilderness had freed me from ordinary party antagonisms. My warnings over the last six years had been so numerous, so detailed, and were now so terribly vindicated, that no one could gainsay me. I could not be reproached either for making the war or with want of preparation for it. I thought I knew a good deal about it all, and I was sure I should not fail. Therefore, although impatient for the morning, I slept soundly and had no need for cheer-ing dreams. Facts are better than dreams."

In the face of almost unspeakable adversity, Churchill boldly addressed Parliament: moving "that this House welcomes the forma-tion of a government representing the united and inflexible resolve of the nation to prosecute the war with Germany to a victorious con-clusion." His words that day still ring with heroism: "I have nothing to offer but blood, toil, tears, and sweat," he said, and predicted "vic-tory . . . victory at all costs."

This time, the French, who had battled the German army for four years in World War I, collapsed in six weeks and it became clear that Britain now stood alone against Nazi aggression. In this dark hour, the nation turned, at last, to the man who had finally ascended to the post he had coveted almost viscerally since the war began: Winston Churchill. And it was his success in the years that followed that for-ever imprinted his image in our minds as the embodiment of the British people: cigar in mouth, cane in hand, bowler on his head, and bulldog tenacity on his face. He became the very symbol of

British stubbornness in the face of crisis, and of her undaunted, even cheerful commitment to liberty in the face of dire challenge.

Few political leaders have lived through so long a period of exile as Churchill had in the decades after his defeat in the Dardanelles. He was seen as a dangerous, anachronistic imperialist, too quick to arms, too slow to negotiations. Too eccentric and outspoken to be trusted, he was excluded from Conservative governments throughout the 1930s. Churchill seemed less the harbinger of the future than an unpleasant reminder of the past.

Was it only Britain's desperation that caused it to turn to him as the war worsened? How did this prophet of gloom, warning constantly of the danger of Germany, suddenly become so stout a repository of optimism that an entire country came to depend on his good spirits?

Part of the answer, of course, is that Churchill was the only leader in British politics who expected war to come and who had urged adequate preparation for it. While all of England seemed content with appeasement, he, almost alone, demanded realism and rearmament—and he, almost alone, was proved right.

But Churchill himself had also changed as the reality around him altered. His militarism no longer seemed a foolhardy preoccupation, born of myopic nostalgia; with the imminent threat before him it took on a new urgency, a new gravitas. And very soon the British people came to see that his strength of character and confidence could be the key to their national preservation. Ultimately, Churchill was propelled into power not only because his warnings about the futility of appeasing Germany before the war were proved accurate, but because he used those eight months to invest Britain with his own optimism, determination, and fortitude.

But it should also be recognized that Churchill's elevation was no accident, nor was it the result of inevitable historical forces. Here was a man who used his position to campaign for the job of prime minister, in the strictest sense, on his merits. He accomplished, in those first eight months of the war, the essential pivot from leader of a faction to leader of a nation, mostly because he saw his opening, recognized that he had what his nation needed, and offered it.

He didn't change his opinions. He simply refocused them, as a camera's lens can be adjusted to take in a wide landscape or the immediate foreground. His goal of maintaining the empire became an urgent commitment to saving democracy. His preoccupation with defense issues now seemed prudent, indeed vital. To the British people his restless mind, once seen as a danger, now emerged as a lifeline, an alternative to stagnation and defeat. His bravado became optimism; his stubbornness, tenacity.

Churchill also took power because he seemed to be the one man who could bring the United States into the war. Throughout the opening of the conflict, Churchill enjoyed the subtle support of President Franklin D. Roosevelt. Though he maintained correct and cordial relations with the Chamberlain government, FDR reached out to Churchill while he was still at the Admiralty.

The two leaders had one thing in common: a love of the sea and of the navy. Roosevelt had begun his national political career by serving as Undersecretary of the Navy at the same time as Churchill ran the British Admiralty in World War I. Capitalizing on their shared passion, Churchill signed his correspondence with Roosevelt "Naval Person"; once he became prime minister, his signature evolved into "Former Naval Person."

The unusual gesture—a direct communication from a head of state to a member of a foreign nation's government without diplomatic portfolio, flattered Churchill, and would help pave the way for American support once he took office.

In power, Churchill rose to the occasion of national leadership as few men or women in history have ever done. No longer the discredited imperialist relic or doomsaying harbinger of war, Churchill stepped boldly forward as the leader of a united and determined country.

At first, though, victory seemed very far away. The German invasion of France had cleverly avoided the feared French defenses at the Maginot Line, executing a flanking movement that isolated the entire British Expeditionary Force and much of the French army in Belgium and northern France. As the Allies fell back to the English Channel they faced annihilation at the hands of the advancing Nazi troops, until rescue came via the British Navy—along with thousands of small boats and pleasure yachts that braved the Channel and German fire to rescue most of the cornered British and French troops. After the evacuation, the threat of German invasion of Britain loomed large. Churchill addressed the nation: "we shall not flag or fail. We shall go on to the end . . . whatever the cost may be, we shall fight on the beaches, we shall fight on the landing grounds, we shall fight in the fields and in the streets, we shall fight in the hills; we shall never surrender . . . until, in God's good time, the New World, with all its power and might, steps forth to the rescue and liberation of the old."

As London faced daily, deadly bombardment, Churchill rallied its people's spirits as no national leader has ever done before or since: "if we fail, then the whole world, including the United States, including all that we have known and cared for, will sink into the abyss of a new Dark Age made more sinister, and perhaps more protracted, by the lights of perverted science. Let us therefore brace ourselves to our duties, and so bear ourselves that, if the British Empire and the Commonwealth last for a thousand years, men will still say 'This was their finest hour.' "

Preparing to accept Churchill's requests for aid, and as a gesture of solidarity, Franklin Delano Roosevelt sent the embattled Churchill a poem by Henry Wadsworth Longfellow that aptly captured the sentiments of the free world:

> *Sail on, O Ship of State!*
> *Sail on, O Union, strong and great!*
> *Humanity with all its fears,*
> *With all the hopes of future years,*
> *Is hanging breathless on thy fate!*

In Reagan and Churchill, we see certain key common elements that allowed each man to come in from the wilderness and lead his nation:

- Each man stuck to his principles, even as the national mood moved away from him. Then, when the pendulum swung back, each was waiting patiently for his time to come.
- Each leader used pessimistic predictions of the outcome of current policies to underscore his proposals for change. But when Reagan's predictions of national gridlock and Churchill's of military disaster were fulfilled, each made the essential pivot

from pessimism to optimism, rallying with confidence and spirit even those voters who had previously rejected his ideology.

- Both figures transcended their ideologies without jettisoning them and drew on their inherent patriotism to carry the day in times of crisis. Reagan's conservatism became "morning again in America," while Churchill's dogmatic militarism became a stand for the civilized world against barbarism.
- Both men were right. Their prophecies proved accurate; their diagnoses of their nation's malaise correct. The essential veracity of their views made it possible for each to wait, as Emerson said, "until this huge world comes round to them."

The war had a similar effect on another nationalist leader—Charles de Gaulle—who entered World War II as an unknown colonel, was hastily promoted to brigadier general, and emerged as a national icon.

DE GAULLE DEFEATS
THE POLITICAL PARTIES

No leader of a democracy since George Washington has had the uncontested political power to shape his country that General Charles de Gaulle enjoyed when the Allies liberated France in 1944. De Gaulle was not merely his nation's leader, hero, or great hope; like Churchill, he became its embodiment. Determined to purge France of the divisive and ineffectual political structure that he believed had led to a humiliating defeat by Hitler in just six weeks in 1940, de Gaulle called for the nation to rise above petty political parties and come together in spirit. Yet it didn't work out that way, and de Gaulle would be forced to retreat and wait more than a decade for his time to come again.

Despite having all the power and prestige of a war hero, de Gaulle was unable to defeat the party system—at least at first. Initiallly, in 1946, when he realized that he had become a Gulliver tied up in procedural knots by parliamentary Lilliputians, he resigned, abrogating all his power, and returned to his quaint country home to lick his wounds. For twelve long years he waited, until he was finally summoned to avert a military coup d'état—or was it to lead one?—in 1958. At last, having regained the power he'd lost, he was able to accomplish his mission—dismantling the political parties that had earlier forced his resignation. De Gaulle had completely abandoned politics rather than renounce his values, had gone into the wilderness, and now returned to witness the vindication of the political principles that had always informed his understanding of the needs of government. And, like Reagan and Churchill, he succeeded because he was able to draw a credible link between his own vision and his people's wellsprings of patriotic feeling.

In exile, de Gaulle—like Reagan and Churchill—had grown. He had gained perspective, reached a deep understanding of his people's need for inspiration. De Gaulle met defeat when he called for procedural change; he found victory when instead he celebrated the once and future grandeur of his nation. In 1946, when he was forced out of office, it was for attacking the indecision and vacillation of the parliamentary system. In 1958, when he returned, it was by speaking with passion of the idea of France.

In other words, the negative critique of France's arcane party system did not succeed—his message fell on closed French ears. But when he switched to a positive articulation of a vision for his nation, he found the French receptive and eager to follow him.

Charles de Gaulle's exalted reputation came from a simple decision he made in June 1940, as France reeled from the massive and successful German invasion. With French and British armies cut off and surrounded, Paris braced for the arrival of the Germans. An obscure colonel and commander of a tank division, de Gaulle had just been promoted to a staff job at the Ministry of War, and was made a brigadier general. Now, amazed, he watched the smoke curling above the city's rooftops as ministries frantically burned state documents.

Unknown to the nation, its army, and most political leaders, France's cabinet was preparing to surrender to Hitler and accept a humiliating peace. Signed in the same abandoned railway car the Allies had used to accept the German surrender in World War I, the treaty left German soldiers to rule half the nation from Paris, and a puppet regime in the town of Vichy to govern the rest. Senile, decrepit Marshal Pétain, the hero of the Battle of Verdun in the previous war, was called to become the Vichy president and dance to the tune of his German masters.

Alone among France's leaders, de Gaulle, a six-foot tower of indignation, set sail for London, where he unilaterally proclaimed himself the leader of a French government in exile. The sheer nerve of the action astonished French, Germans, British, and Americans alike.

With no mandate from France's elected leaders, the National Assembly, the cabinet, or even his fellow army officers, this mid-level functionary had stepped into the breach and declared himself the leader—and savior—of a storied nation that appeared to be nearing its final hours.

As de Gaulle wrote in his memoirs, "Deliberation is the work of many men. Action, of one alone."

As a three-line headline blared across the top of the *New York Times* of June 23, 1940—FRENCH SIGN REICH TRUCE—a brief item appeared near the bottom of the page: "General Summons French To Resist." Only in the subhead was the name of the unknown and presumptuous leader mentioned: "De Gaulle Offers to Organize Fight Abroad." From London, the story informed readers that this lone wolf had broadcast a message to the French people, "calling on them to continue the fight against Germany by every means in their power."

Even to a world desperate for good news, it seemed like a minor anecdote, a sidebar to a disaster of monumental proportions. In truth, London was crowded with governments-in-exile, as Hitler subdued virtually every nation on the European continent.

But de Gaulle was different. Not only was he speaking for France, but he actually had an army to lead—thanks to Churchill's skillful evacuation of both French and British troops across the Channel as France was falling. At a conference in France, Churchill had promised Premier Paul Reynaud that he would rescue French troops along with his own; Churchill linked arms with the Frenchman and paraded around a table, yelling—in French—"arm in arm, arm in arm, arm in arm" to describe how the evacuation of the two armies would proceed. The British rescuers, including many pleasure craft, succeeded in evacuating 139,911 French from a narrow beachhead at Dunkirk, as well as 338,226 British soldiers. These French soldiers would become de Gaulle's army; and, with Churchill's all-important backing, the new commander insisted on representing his defeated nation in the top councils of the Allied leaders.

De Gaulle's audacity was breathtaking. Using his new British- and American-equipped army, he seized the islands of St. Pierre and Miquelon and declared them liberated in the name of Free France. What made this bold stroke less than popular among the Allies was the unfortunate fact that these islands lay not in German hands, but off the coast of Canada! Like a confused football player, de Gaulle seemed to be sprinting for a touchdown in the wrong direction.

In French North Africa, de Gaulle proved himself a nuisance to Roosevelt, Churchill, and Eisenhower as he killed their plans to install a puppet leader, Admiral Jean François Darlan, in territory their troops had seized from the retreating Germans and Italians. Epitomizing Henry Kissinger's observation that "the weak gain strength through effrontery," de Gaulle inspired French men and women everywhere with his insistent demand that France regain her position in the sun. And in his wake resistance rallied, in France itself and in her colonies. "As long as the war lasted," de Gaulle later recalled, "I had the means, morally speaking, to muster the French people."

Churchill was indulgent toward de Gaulle's usurpation of power, but FDR was not. Angered by the Frenchman's arrogance, Roosevelt was eager to undermine his position as leader of the Free French. But Churchill recognized the general's importance. "You will remain the only choice," he counseled de Gaulle in 1942. "Don't confront the Americans head on. Be patient! They will come to you, for there is no other alternative."

As Free French forces joined British, American, and Canadian units in the liberation of their land, de Gaulle's prestige soared. Marching tall at the head of his troops, he paraded up the Champs-Élysées in August 1944, passing through the Arc de Triomphe as Parisians erupted with joy after four years of brutal Nazi occupation. Brushing aside Roosevelt's anger, the former colonel proved Churchill right: After the war he emerged as the only choice to head the new government in newly liberated Paris.

But one question haunted postwar France: What had gone wrong? How had the nation grown so weak that it could fall so swiftly to the German onslaught of 1940? France had been one of the most powerful nations in the world. During World War I it had resisted the might of the German army for four years, and then triumphed over it, but in World War II the nation was so rotted from within that it succumbed in just six weeks to Hitler's forces.

De Gaulle attacked their "weakness," and wrote that their "decadence" was "still concealed beneath rhetoric," "no longer inspired by principles." He predicted that they tended toward "degradation, shrinking until each became nothing more than the representation of a category of interests." Turning power over to the parties, he charged, "could result only in impotence."

Now that the old Third Republic, dominated by the political parties, had fallen, de Gaulle argued that it should be replaced by a radically different form of government—one that dispensed with parties altogether. "If the government fell into [the parties] hands again," he wrote later, "it was certain that their leaders, their delegates and their militant members would turn into professionals making a career out of politics. The conquest of public functions, of influential positions, of administrative sinecures would henceforth absorb the parties and limit their activities to what they called tactics, which was nothing more than the practice of compromise and denial."

De Gaulle demanded that France rise above the party system. In 1945, addressing the Paris Municipal Council, de Gaulle contended that his nation was "discovering with a clear mind the efforts she must make in order to repair what this war . . . has destroyed of her substance." He issued a stark challenge to his people: "We will re-establish ourselves only by arduous labor, a severe national discipline . . . Let the rivalry of parties be silenced!"

De Gaulle later explained: "As I saw it, the state must have a head, that is, a leader in whom the nation could see beyond its own fluctuations, a man in charge of essential matters and the guarantor of its fate." Not surprisingly, it was a description he felt tailor-made to fill. "As the champion of France rather than of any class or party, I incited hatred against no one and had no clientele who favored me in order to be favored in return." But he refused to be a despot. He would rule a democracy, or he would not rule at all. "Even as I dismissed the notion of my own despotism, I was no less convinced that the nation required a regime whose power would be strong and continuous. The parties were evidently unqualified to provide such power."

How could a democracy function without parties? De Gaulle's answer was to propose a new constitution for France in which the chief executive would not be chosen by a parliamentary party, but would "serve only the national community." The president must "not belong to a party," he charged; the only thing a political party was good for was "cooking its own little soup on its own little fire, in its own little corner." Instead the president should be "designated by the people, empowered to appoint the Cabinet, [and] possess the right to consult the nation, either by referendum or by election of assemblies."

But the political parties, which failed to see quite the same merit in their own destruction, would not agree to go quietly. Of course, de Gaulle's "dominance of the French scene was to all appearances as great as ever" in 1945, as biographer Don Cook writes. "He had only to rise in the Constituent Assembly for it to come to attention, like an unruly schoolroom. . . . The trouble was that when the teacher left, the class began throwing spitballs at each other all over again."

De Gaulle was never a natural parliamentary politician. He did not like the "parliamentary sapping of the roots and trunk of power," in Cook's phrase. Yet the parties were coming back, and were "now rapidly overtaking the lone supremacy he had enjoyed when he walked down the avenue des Champs Élysées the year before."

De Gaulle was aware that the parties were working to frustrate his agenda. "I could not overlook the fact that my project contradicted the claims of every party. . . . It was clear that in the crucial debate that was approaching, discord was inevitable. With various qualifications, all the parties intended the future constitution to re-create a regime whose powers would depend directly and exclusively on themselves and in which de Gaulle would have no place unless he were willing to be merely a figurehead."

In October, 1945, France held its first national election since liberation. As Cook observed, "political activity was growing steadily more complex and intense. France was rapidly returning to what de Gaulle called 'the game of the parties.' " Yet de Gaulle had no party to represent his views. Contemptuous of the very idea, he hoped to govern by referendum and plebiscite, consulting the people directly on the policies he chose to pursue. In military terms, lacking the ground troops—a party of supporters—to take parliamentary seats, de Gaulle had only the air power of his own soaring prestige to impose his will.

When the time came, de Gaulle was obliged to watch as his partisan adversaries carved up the seats in the legislature without him. The Communists, his nemeses, emerged as the strongest party in France, with 26 percent of the vote, trailed by the Socialists with 25 percent. The MRP Christian Democrats—the party de Gaulle disliked the least—got only 26 percent.

Though his enemies had a majority, the new Assembly could not ignore de Gaulle's status as national leader, and in November 1945 they duly elected him president. But even as he took office, the proud general realized just how fiercely he would have to compete with the parties for political power. As soon as they elected him, the Chamber of Deputies shackled de Gaulle by specifying that he could not chair the Council of Ministers or the Committee of National Defense, and restricting his power to grant pardons—particularly important as the war crimes trials approached.

Everyone knew the new president could never abide such restrictions. In the words of historian Jean Lacouture, it was like trying "to force him into a uniform that would have been too tight for a local mayor."

Not a man to suffer indignity silently, de Gaulle returned from a brooding vacation on the Côte d'Azur on January 14, 1946, and decided to resign. It was a bluff, a tactic to win approval for his vision of a new France. "Before a week is up they'll be sending a delegation asking me to come back," he said.

De Gaulle called a cabinet meeting for the following Sunday, January 20. "The exclusive regime of the parties has come back," he told them. "I disapprove of it. But short of establishing by force a dictatorship, which I don't want and which would probably turn out badly, I lack the means to prevent this experiment. I must therefore retire. This day, in fact, I shall address to the president of the National Assembly a letter informing him of the government's resignation." With that, Cook noted, "he nodded a farewell to all and walked out."

De Gaulle's resignation was, in a sense, no surprise; it reflected nothing more than his inability to reconcile his own sense of grandeur with the daily realities of party politics. Yet he seems never to have considered that he might not be asked back quickly. Comparing himself to France's patron saint—de Gaulle was never short of ego—he mused, "Really, one can scarcely imagine Joan of Arc married, the mother of a family and, who knows, with a husband unfaithful to her."

But the ploy backfired. De Gaulle listened, but in vain, for the footsteps of a messenger asking him to return. As he told his nephew a few years later, "I have made at least one political mistake in my life:

my departure in January 1946. I thought the French would recall me very quickly. Because they didn't do so, France wasted several years."

And yet there's a lesson in de Gaulle's departure that should be noted by anyone—in politics or business—faced with an irreversible shift in fortune: Rather than presiding over a process that would inevitably have eroded his reputation, he opted "to withdraw from events before they withdrew from me," and left, musing, "I prefer my legend to power." Rather than clinging to position, he preserved his personal standing by stepping aside. And though his next twelve years were spent in the wings, by his own reckoning they were not years he wasted.

A keen strategist, de Gaulle realized that he needed parliamentary ground troops in his political wars—warm bodies to put into the seats of Parliament, who would vote to return him to power and relinquish their own in the process. And so in 1947 he established a new political party, the RPF (Rally of the French People). It was hardly a conventional party: De Gaulle declared, in the founding statement of the RPF, that its object was "to promote, above the parties, in economic, social, imperial and foreign affairs, the solutions I have indicated." It was, in other words, the party to end all parties.

André Malraux, the chief party spokesman, differentiated what came to be known as "Gaullism" from the beliefs of other parties, saying that while "every group, party and union and association acts and speaks for itself, as if it were independent of the rest," the RPF had "declared that the party system in France, as it functions at the present time, is in no condition to take measures for the public welfare."

At first it appeared that the RPF would sweep to power. In the municipal elections of 1947 it won 40 percent of the vote, carrying thirteen of the top twenty-five cities of France, including the two largest, Paris and Marseille. But by the next election—the parliamentary balloting of 1951—the ardor for the RPF had cooled, and it received only 23 percent of the vote.

Chastened, de Gaulle publicly admitted failure. "The efforts that I have put in since the war . . . have not so far succeeded. I do not deny it. This is, it must be feared, to the detriment of France." And when he left office, de Gaulle now seemed to leave politics entirely. Like Reagan after Ford beat him, like Churchill in his fall from grace in the 1920s, he went down fighting for principle.

Yet de Gaulle used his years in the wilderness to refashion his approach and refine his theme. As his biographer Jean Lacouture wrote, "we must not confuse the de Gaulle of 1945 with the polished de Gaulle of 1958 and after." A new strain began to emerge in de Gaulle's rhetoric during his years out of power, as he awaited the summons to lead again.

As he moved among the French people—from whom he had been removed during the wartime years—he began to evolve a new sense of his mission. "It was during those years of travelling through France, of endless tours, of visits to ordinary people, of nights spent in their homes, that de Gaulle learnt to know the French," confidant Pierre Lefranc remembered. "He acquired a familiarity with the people that is the key to his behaviour after 1948, of his art of finding the right tone of voice, the striking arguments, a simple way of talking, the art of persuading. For de Gaulle the [campaign for the] RPF had been his discovery of the French."

Where once he spoke for himself in opposing the parties, now de Gaulle began to learn how to speak for the French in demanding national greatness. Writing his memoirs while in the wilderness, de Gaulle seemed to find his voice. There was, according to *Gaullism* author Anthony Hartley, "a strong element of romanticism, not to say mysticism, in de Gaulle's concept of his role as sort of vicar on earth of an eternal and unchanging France. To hold such a view of his own political leadership required an overwhelming sense of mission and a boundless confidence in the correctness of his own interpretation of the transcendent value which had chosen him as its temporal representative."

De Gaulle's wartime memoirs, finally published after his return to power in 1960, reflect this new twist on his former views. "The whole atmosphere of the *Mémoires* is messianic," Hartley writes. "It is the tale of a great national leader, like Moses, guiding his people back out of the wilderness into the Promised Land that had once been theirs."

Just as Ronald Reagan began to shift his focus from conservatism to the destiny of America, and Churchill turned his attention from preserving the empire to saving Britain and civilization, so de Gaulle began to speak of the grandeur of the French nation. In the 1940s, de Gaulle had seemed most motivated by what he opposed—the system of partisan government. In the 1950s, he began to focus on what he valued—the greatness of France. Rising above procedural issues, he increasingly described himself not as an opponent of parties, but as an advocate of France. By the time his transformation was complete, he had fulfilled a prediction he made back in 1947: "The day will come when, rejecting sterile games and reforming the ill-built structure in which the nation has lost its way and the state has lost all authority, the mass of French people shall gather together upon France."

During the years de Gaulle spent licking his wounds in exile, the new French government floundered indeed. In turmoil, the parliamentary system triggered constant changes in government. In its twelve miserable years, the Fourth Republic saw twenty-six new regimes. Prime ministers changed more than once a year, their weak and fragile coalitions collapsing beneath them, unable to accomplish anything on either the foreign or the domestic front.

De Gaulle pointed out that self-interest dominated the Fourth Republic's politicians more than it normally would, because the rapid turnover in prime ministers and governments meant that they did not have the political security to take action. The less they did, the more likely they were to survive. What Americans now call "gridlock" the French called "immobilisme."

Part of the problem was that about a third of the members of the Chamber of Deputies were either Communists or Gaullists, opposed ideologically not just to the government but to the system. Any would-be prime minister had to attract his coalition majority from the rest, ensuring government instability.

The Fourth Republic ended and de Gaulle took power in 1958, largely because of a protracted and bloody war to keep Algeria in the dwindling French Empire. Having lost its colonies in Indochina and its other possessions in North Africa, France was determined to keep Algeria. Colonial advocates like Jacques Soustelle, who became governor-general of Algeria in 1955, believed that it was precisely because of these other reverses that France needed to hold onto Algeria.

With one million French men and women living in Algeria amid a native population of nine million, the stakes for France were quite high. As nationalist rebels fought a war for independence, the dictatorial French regime reacted brutally, launching a campaign of tor-

ture and terror to flush out the rebels. France considered Algeria
not a colony, but a province that elected delegates to the national
parliament. The French who lived in Algeria clung passionately to
Paris for fear that their lives, rights, and property would be lost if
Algeria should become independent. Called the *pieds noirs* (black
feet) because they wore black shoes while natives customarily went
barefoot, they grew increasingly dissatisfied with the weak regimes of
the Fourth Republic. Demanding strong action to keep Algeria
French, they organized a massive strike on May 13, 1958. Supported
by the French Army stationed in Algeria, they took over the
governor-general's office and set up a committee of public safety to
rule under direct military control. Desperate to restore strong leadership to the national French government, they called for de Gaulle
to return to power. Soon hardened French paratroopers, based in
Algeria, were threatening to descend on Paris and install de Gaulle in
office.

"This indeed was the moment or the opportunity," Don Cook
wrote, "de Gaulle had been waiting for." On May 15, 1958, de Gaulle
issued a statement signaling his willingness to take power. It was, he
told the French people, a matter of national preservation. "The
degradation of the State inevitably brings with it the distancing of the
associated peoples [meaning Algeria], disturbance in the fighting
forces, national dislocation, loss of independence. For twelve years,
France, in the grip of problems too severe to be solved by the regime
of the parties, has embarked on this disastrous process.

"Not so long ago, the country, in its depths, trusted me to lead it,
in its entirety, to its salvation. Today, with the trials that face it once
again, let it know that I am ready to assume the powers of the
Republic."

In this new political incarnation, de Gaulle sublimated his condemnation of parties to focus instead on the "trials" France faced,
"the national dislocation" and the "loss of independence." Like Reagan challenging America to rediscover optimism at the end of the
Carter years or Churchill summoning the British people at their
finest hour, de Gaulle spoke not about ideology or reform but on his
country's need to save itself.

"On May 19 de Gaulle drove to Paris to hold a news conference—
probably the most masterful and dramatic of his his long life. It was

his first meeting with the press in three years . . ." Addressing thirteen hundred journalists, he discussed the Algerian crisis, but proposed no specific solution. He said only that he would not "specify at
the present time what the conclusions of my arbitration would
be . . ." It was the statement of a wise village elder, not a campaigning
firebrand: "I know of no judge who hands down his decision before
hearing the case." And he left the stage with a gesture that left his
audience wanting more: "Now I shall return to my village and I shall
remain there at the disposal of the country." The public reaction was
ecstatic.

A week later, de Gaulle returned to Paris to meet with the ineffectual and frustrated Fourth Republic premier Pierre Pflimlin, who had
been in office only fifteen days. De Gaulle's goal, he said, was to
ensure "the unity and independence of the country." Although his
new ascent to power was clearly impelled by the widespread fear of a
military coup, he stressed that he was setting in motion "the regular
process necessary for the establishment of a Republican government.
I believe this process will be continued and the country will show, by
its calm and its dignity, that it wishes to succeed."

"Any action endangering public order," he warned, "from whatever side it originates, could have grave consequences. . . . I could
not give my approval."

France's president, René Coty, and Pflimlin saw that they had no
choice but to accede to de Gaulle. Both realized that if they failed to
hand the government over to de Gaulle, the paratroops from Algeria
would arrive to do it themselves. On May 28, Pflimlin resigned. President Coty told the National Assembly, "I have called on the most
illustrious of Frenchmen, who, during the darkest years of our history, was our leader for the conquest of liberty and who, having
achieved national unanimity around his person, spurned dictatorship, to establish the Republic."

At the heart of this transfer of power was the French people's
deep fear of armed revolt. And de Gaulle, the storied national icon,
was the one figure behind whom all sides could unite. Lacouture
noted that one French political leader said, "The Gaullist strategy
consisted of allowing the political class to see the threat of violence
and therefore to side with [de Gaulle], and to let the soldiers believe
that he was their man." Jean Chauvel, a former French ambassador

to Britain, put it best: "All liberties being in danger, de Gaulle offered the only possibility of saving some of them."

Yet the de Gaulle who obsessed over the niceties of constitutional reform in 1946 could scarcely have given such assurance to the military on the one hand and the French people on the other. It was the new Charles de Gaulle, who spoke of the grandeur of France and the needs of the nation, whose message resonated with every French citizen. In subsuming his demands for reform into a broader appeal to save the nation, de Gaulle had sealed his own return to power.

On June 1, 1958, de Gaulle appeared in person before the National Assembly and demanded full emergency powers for six months; in the interim, the legislature would take a vacation "and he would govern by decree." He wanted "a mandate to draw up a new constitution for a Fifth Republic that would then be submitted for approval by national referendum"—his favored political tool of old. With no other reasonable options, the Assembly agreed to give de Gaulle what he wanted by 329 to 224, with 32 abstaining. Most Communists and many Socialists voted no, among them the future Fifth Republic president François Mitterrand.

De Gaulle traveled into Algiers on June 4, 1958, with the same aura of command he had shown on entering Paris in 1944. It was "liberation all over again," Cook wrote. Addressing a largely European crowd, de Gaulle carried a message of reassurance. "I have understood you," he said, and "spoke of renewal, fraternity, reconciliation and all Frenchmen voting together for a new constitution." It was a virtuoso performance, and upon his return to Paris it became clear that he would get exactly the constitution he wanted.

The Fifth Republic invested the president with strong national powers to run the government, including sole control over foreign policy and the military. He alone could select and dismiss prime ministers (although they still required the support of a majority in the National Assembly). His term of seven years was absolute; he could not be removed from office. He could call national plebiscites or direct referenda at will. Initially the president was to be chosen by an electoral college of leading Frenchmen. By 1962, France amended the constitution to provide for the direct election of the president.

In the first legislative elections the Gaullist Union pour la Nouvelle République (UNR) won more than 200 of the 465 Assembly seats. This ensured "almost monolithic political power" for de Gaulle. Finally, on December 21, 1958, the grand council of electors installed de Gaulle as the first president of the Fifth Republic—with 78 percent of the vote.

In the end, de Gaulle gave Algeria her freedom and dismantled virtually all France's empire. Sputtering in impotent rage, the pieds noirs fled to France and the extremists set about conspiring to arrange de Gaulle's assassination. They were not successful. De Gaulle's presidency lasted until 1969; the Fifth Republic lasts to this day and shows every sign of continuing.

It was a fascinating vindication of this maverick military hero's political instincts. Ousted from power by a nation unprepared for the political changes he found necessary, de Gaulle left the stage; let the flawed political system run its course; and then, when the inevitable crisis loomed, charged forward to save the day—in the name of the country he had rescued once before. And by framing his reemergence in the noblest patriotic terms, he found he was able to secure the very political reforms his people had once rejected. Long-delayed political reforms, accomplished in a time of national crisis. It was a story that would have been familiar to another figure who would become a symbol of his country—Abraham Lincoln.

EXAMPLE FIVE—SUCCESSFUL

ABRAHAM LINCOLN MOVES FROM ABOLITIONISM TO UNION . . . AND WINS

Though Abraham Lincoln was personally opposed to the institution of slavery, both before and after his election as president in 1861, he consistently denied that he sought to abolish the practice, or free the legions of slaves held by white citizens throughout the slaveholding South. He was, he said, only interested in stopping the institution's spread to new territories and states. When the Emancipation Proclamation took effect, during wartime, on January 1, 1863, opposition Democrats were quick to point out the inconsistency in his positions on the slavery issue. Mockingly, tauntingly, they sang:

> *Honest Abe, when the war began*
> *Denied abolition was part of his plan*
> *Honest old Abe had since made a decree*
> *The war must go on til the slaves are all free*
> *If both can't be honest, will someone tell me how,*
> *If honest Abe then, he is honest Abe now?*

They had a point. While Lincoln never wavered in his personal condemnation of slavery, and his public political position never changed, his rhetoric—and focus—passed through four distinct phases until he was able to deal a mighty blow for freedom. Just as

Reagan, Churchill, and de Gaulle began their careers by focusing on their ideological objective, assuming a broader, more patriotic rhetoric only after defeat, so Lincoln also spent his time in the wilderness. Indeed, like the others, Lincoln was able to achieve his longstanding personal goals only when a climate of national peril seemed to transform the question from a factional dispute into a matter central to the preservation of the Union itself. His stand on principle was vindicated only when it became part of a larger national objective.

Lincoln's entry into politics had little or nothing to do with the slavery issue. A member of the Whig party, he represented his district in the Illinois legislature by demanding increased state spending on internal improvements. Echoing the American System, a plan advocated by Whig Party leader Henry Clay, Lincoln backed a combination of public works and high tariffs to promote industrial and economic growth.

Abraham Lincoln served one term in Congress, from 1846 to 1848, before choosing to honor a campaign pledge to step down and give another deserving Illinois Whig a chance to serve. During his years in Congress, he focused primarily on opposing what he saw as American imperialism and expansionism in the Mexican War. While President James K. Polk maintained that the war began when American troops were assaulted by Mexicans on U.S. territory, closer examination revealed that the incident took place on disputed land in Texas that both nations claimed as their own. Less than popular for his opposition to the war, Lincoln left public life for a lucrative law practice as a utility lawyer.

When he reentered the public arena in the late 1850s, it was as a strident opponent of the expansion of slavery into new states and territories. With the Whig Party dying, Lincoln worked to help establish the new Free Soil Republican Party. Denouncing slavery with moral and religious fervor, Lincoln "was considered . . . to be closer to the radical than to the conservative wing of the Republican Party," sharing "the radicals' moral abhorrence of slavery and was fully committed to its eventual eradication." As Lincoln told a Chicago gathering in 1858, "I have always hated slavery, I think as much as any abolitionist." Later he told a Wisconsin audience that the Republican Party was defined by "hatred to the institution of slavery; hatred to it in all its aspects, moral, social and political."

While Lincoln inveighed against the moral evils of slavery, he was careful not to identify himself with the goals of the abolitionist movement. Instead, Lincoln focused on stopping the spread of slavery to new territories as they became states of the union. As the *New York Times* observed, Lincoln believed that "the barbaric institution of slavery will become more and more odious to the northern people because it will become more and more plain . . . that the States which cling to Slavery thrust back the American idea, and reject the influences of the Union." And his rhetoric was passionate and memorable; speaking of the "ultimate extinction" of slavery in 1858, he declared that "a house divided against itself cannot stand."

When Lincoln challenged Stephen A. Douglas, the incumbent Illinois senator, in 1858, he focused relentlessly on the evil that slavery represented. In the candidates' historic debates that year, the moral validity of the "peculiar institution" became the central ideological debate between the two men.

Yet ideology alone wasn't strong enough to carry the day. Although Lincoln received more votes than Douglas, the incumbent won enough votes from holdover members of the state legislature— who actually chose the senator—to guarantee his reelection.

As with Reagan, Churchill, and de Gaulle it was after defeat that Lincoln's rhetoric began to change. As the friction over slavery began to consume the nation—and talk of secession spread throughout the South—it soon became clear that America's future would rest on the resolution of this difficult issue. And as he contemplated his first run for national office, these two goals—limits on the expansion of slavery, and the preservation of the Union— now became co-equal and inseparable in Lincoln's speeches. He was entering his second rhetorical phase.

Lincoln realized that the Republican Party had lost its first bid for the White House in 1856 by losing the key northern states of New Jersey, Pennsylvania, Indiana, and Illinois. To win in 1860, the party would have to moderate its image to carry these pivotal states. To do so, the Republican Party shifted its focus away from the slavery issue. For the Republican Party of 1860, the real priority was the survival of the Union—whatever the threat.

Indeed, according to historian Eric Foner, "one of the major reasons many conservatives supported Lincoln in 1860 was their belief that his election would preserve the Union by finally ending the slavery controversy. While the conservatives would sacrifice [their] anti slavery [position] to save the Union, and the radicals would endanger the Union to attack slavery, the moderates, with Lincoln at their head, refused to abandon either of their twin goals—free soil and Union.

"Lincoln agreed with the Webster/Clay tradition when he insisted that the Union preceded the Constitution and was a creation of the American people, not a compact between states." Foner continues, "But to their devotion to the Union as the paramount end of politics he added the radical conception of the Union as a means to freedom. To preserve the Union by undermining this purpose would be to subvert the foundations of the Union itself. The goals of the Union and free soil were intertwined, and neither could be sacrificed without endangering the other."

With this equation of the Union with antislavery values, Lincoln merged his opposition to slavery into the larger ideal of patriotism and national purpose. "The Republican position on the Union as it emerged in the secession crisis was that the Union should be revered and defended not only for itself," Foner writes, "but also because of the purposes for which it had been created. High among these purposes was the spread of freedom which in the 1850s meant the confinement of slavery."

At Cooper Union in New York City, on February 27, 1860, Lincoln gave an address that reframed his position, and that of his party, in far less radical terms than in years past. The Republican platform, he told the audience, was merely a moderate attempt to preserve the Founders' original vision of the country in the face of radical threats from the slaveholding states. Lincoln took great care to use George Washington's cloak to shelter his opposition to the expansion of slavery, noting that the first president had signed the Northwest Ordinance prohibiting the expansion of slavery into the region. "George Washington . . . was then President of the United States, and as such approved and signed the bill," he reminded the crowd, "thus completing its validity as a law, and thus showing that, in his understanding, no line dividing local from federal authority, nor anything in the Constitution, forbade the Federal Government, to control as to slavery in federal territory."

Indeed, Lincoln spent much more time defending the Union than attacking slavery in his Cooper Union address. Pointing a finger at those "who will break up the Union," Lincoln addressed southern secessionists directly: "Your purpose, then, plainly stated, is that you will destroy the Government, unless you be allowed to construe and enforce the Constitution as you please, on all points in dispute between you and us. You will rule or ruin in all events." His language and imagery grew more direct as he continued.

"Do you really feel yourselves justified to break up this Government?" he asked the South in absentia. Referring to southern threats to leave the Union if the Democrats lost the 1860 election, Lincoln said, "But you will not abide the election of a Republican president! In that supposed event, you say, you will destroy the Union; and then, you say, the great crime of having destroyed it will be upon us! That is cool. A highwayman holds a pistol to my ear, and mutters through his teeth 'Stand and deliver, or I shall kill you, and then you will be a murderer!' "

As with Reagan, Churchill, and de Gaulle, the national environment was deteriorating around Lincoln as he moved ahead with his political career. The closer he got to the White House, the more likely it seemed that the Union would dissolve before his eyes. As he grew from a little-known, dark-horse candidate to the Republican nominee in 1860, the threat of southern secession became more and more real.

By the time of his election in November 1860, Lincoln faced the strong possibility of an imminent split in the Union. Now, once again, his rhetoric entered a new phase—its third—moving yet farther away from the basic issue of slavery and toward the need to keep the federal union intact.

In his inaugural address of March 4, 1861—as the South was leaving the Union—the new president spoke very little about the basic question of human freedom. Indeed he denied any interest in freeing the slaves, declaring that he had "no purpose, directly or indirectly, to interfere with the institution of slavery in the states where it exists."

Was he honest in this forthright declaration? Historians have debated the point, but there's no contesting the fact that by the time he took office the need to save the Union had eclipsed all other goals in his mind. Now he recognized that he would have to call upon the people's deepest feelings for their country if he wanted to block secession. "Though passion may have strained," he reminded listeners, "it must not break our bonds of affection. The mystic chords of memory . . . will yet swell the chorus of the Union." And he asked Americans to step back and reexamine the roots of their burgeoning conflict: "Before entering upon . . . the destruction of our national fabric . . . would it not be wise to ascertain precisely why we do it? Will you hazard so desperate a step while there is any possibility that any portion of the ills you fly from have no real existence?"

After reviewing the legal case for an indivisible Union, Lincoln returned to more fundamental truths. "Physically speaking, we cannot separate. We cannot remove our respective sections from each other nor build an impassable wall between them. A husband and wife may be divorced and go out of the presence and beyond the reach of each other, but the different parts of our country cannot do this."

With a new civil war upon him, Lincoln became totally absorbed in the maintenance of the Union—just as Churchill and de Gaulle became focused entirely on the preservation of their nations. At first, Lincoln had spoken defiantly and clearly against the moral evil of slavery. Then he began to stress the need for union even as he decried slavery. As he was inaugurated, he appealed to the South to remain in the Union, and largely avoided even mentioning the divisive issue that had brought the nation to the brink of civil war. Now his rhetoric and positioning entered a fourth and final phase, as he assumed the awesome responsibilities of the presidency amid the unprecedented challenges of Civil War.

Recognizing that his first task as president was to keep slaveholding border states like Kentucky, Missouri, Maryland, and Delaware in the Union, Lincoln was forced to take stands that seemed to fly in the face of his old priorities. When John C. Frémont, the 1856 Republican nominee for president and now a Union general, ordered slaves in Missouri emancipated, Lincoln angrily rescinded the order. His opposition to slavery had been overwhelmed by the paramount goal of saving the Union.

When radical Republican newspaper editor Horace Greeley wrote an editorial titled "The Prayer of Twenty Million" that pleaded with Lincoln for emancipation in August 1862—after the end of the war's first bloody and brutal year—Lincoln replied in a famous letter that made his priorities quite clear. "If I could save the Union without freeing any slave I would do it, and if I could save it by freeing all the slaves I would do it; and if I could save it by freeing some and leaving others alone I would also do that. What I do about slavery, and the colored race, I do because I believe it helps to save the Union; and what I forbear, I forbear because I do not believe it would help to save the Union."

". . . If there be those who would not save the Union, unless they could at the same time *save* slavery, I do not agree with them. If there be those who would not save the Union unless they could at the same time *destroy* slavery, I do not agree with them. My paramount object in this struggle *is* to save the Union, and is *not* either to save or to destroy slavery."

Even his ultimate stand against evil—the historic Emancipation Proclamation—Lincoln explained on grounds not of morality or freedom, but of strict military necessity.

First, he needed to prevent Britain—whose economic interests were intertwined with the cotton trade of the South—from recognizing the Confederacy. Knowing that British voters would not tolerate intervention on the side of slavery, Lincoln felt it necessary to clarify the war's basic issue.

He also needed black troops to fight for the Union. As the war became ever more bloody, and recruitment became harder in northern cities, he looked to the liberated slaves as a source of manpower. Perhaps even more, he wanted to lead their enslaved brethren to escape and cripple the agricultural base of the South, whose white farmers had left their fields in the hands of their slaves to fight in the war.

Now, even as Lincoln achieved the goal of his lifetime, he felt constrained to spin it as simply a pragmatic step to help win the war. How far he had come from the man who, only five years earlier, had said that the nation "cannot endure, permanently half slave and half free." Like de Gaulle, who refrained from attacking parties as he came to power to protect France; Reagan, who sublimated his conservatism in the ecumenical spirit of bringing optimism back to America; and Churchill, who dropped the rhetoric of empire while the Nazis threatened his nation's very survival, Lincoln changed his message, and that change allowed him to succeed.

In the great sweep of history, leaders are often confronted with this need to reposition and reframe their idealism and ideologies in the broader context of national renewal. The lesson is twofold: first, it bespeaks a people's basic hesitation to consider substantial change—unless and until the status quo simply seems untenable. Second, and perhaps more important, it suggests just how vital it is to remind constituents of the greater context of the debate—of the shared beliefs that hold them together. If a leader can make a convincing case that his people's essential values are part of what's at issue, he or she is far more likely to prevail.

When one seeks radical change without making an appeal to national purpose and patriotism, on the other hand, the results can be disastrous. Woodrow Wilson, in his crusade to persuade the United States to join the League of Nations—which he himself had established—is an example of a principled stand that failed to carry the day. And the disaster was due, in large part, to his failure to draw a persuasive link between his cause and the future of his nation.

EXAMPLE SIX—UNSUCCESSFUL

WOODROW WILSON GOES DOWN FIGHTING FOR THE LEAGUE OF NATIONS

To understand fully how important it was for leaders like Reagan, Churchill, de Gaulle, and Lincoln to morph their ideologies into a broader national and patriotic vision in order to get them adopted, it is instructive to examine the case of one who refused to take such steps. Woodrow Wilson's vision of a League of Nations went down to historic and tragic defeat precisely because he failed to make the transformation that each of these other world leaders made during their years in the wilderness.

No twentieth-century American president suffered as major a foreign-policy blow in Congress as Woodrow Wilson endured when the Senate refused to approve U.S. entry into the League of Nations—it was a move that destroyed both his presidency and the man, leaving him broken and ill.

Wilson's sponsorship of the League was clearly right-minded, his opponents' position obviously wrong. The disastrous interwar years, and the renewal of the conflict between Germany and the Western Allies after a bare twenty-one years of so-called peace, make it evident. And the success of the diplomatic matrix of international cooperation—including the United Nations—that has evolved since World War II fully vindicates Wilson's vision.

Woodrow Wilson should have prevailed. He entered the battle at the summit of his popularity, having led the nation to a global victory in war. It was he, more than any other man, who had ended World War I by extending a generous fourteen-point peace proposal to the German government. Hailed around the globe, Wilson was destined to be a prophet without honor in his own land, as his proposed League went down to a humiliating defeat.

Wilson had entered politics only a few years before he ran for president. As president of Princeton University, he showed a commitment to reform and good government that catapulted him first to the governorship of New Jersey and later, in 1912, to the presidency of the United States. He succeeded by riding the enormous wave of enthusiasm for reform kindled during the years of Theodore Roosevelt's presidency (1901-08). Roosevelt had slashed away at the alliance of corrupt politicians and ruthless businessmen that had largely run America in the years since the Civil War. America's first reform president, he passed tough standards for the infamous meat-packing industry, antitrust legislation to curb monopolies, public regulation of railroads, conservation of national parks, and a host of progressive measures. But, when his designated successor, William Howard Taft, seemed to backtrack on his reforms, Roosevelt was livid. Returning from a postpresidential African safari he became determined to hunt down and replace Taft.

Elected in an upset when Taft and Roosevelt split the Republican vote, Wilson, as a reformer, had the wind at his back. An eloquent speaker, he sold a national program of reform that was second only to Roosevelt's own achievements. In his first term, Wilson amassed a formidable record of progressive change. Enacting an agenda he called the New Freedoms, he established the Federal Reserve Board to regulate banks, strengthened antitrust laws, passed woman's suffrage, the income tax, direct election of U.S. Senators, and, less wisely, enacted prohibition of the sale of alcoholic beverages.

The first president to go personally to Congress to deliver his State of the Union address, Wilson understood how to pluck the chords of national emotion to rally supporters to his cause. Indeed, his ability to read the minds of the American people in his early years as president rivals that of our greatest chief executives.

When World War I broke out in 1914, Wilson was determined to keep America out of the fight. Reelected, barely, on the slogan "he kept us out of war," Wilson watched in horror as the war in Europe ground into dust the best of a generation.

But America's entrance into the war became inevitable when Germany announced a policy of unrestricted submarine warfare against U.S. ships in commercial trade with Britain and France. Knowing that this policy would force the United States into the war, the German kaiser gambled that the blockade and his troops in France could defeat the Allies before American forces could make themselves felt.

The resulting destruction of U.S. merchant ships forced Wilson's hand as he led the nation into war. Vowing to "make the world safe for democracy," Wilson summoned national spirit to a new standard of idealism, calling for a "war to end all war."

Wilson rallied America with his eloquence. Summoning a patriotic fervor, he mobilized the country and helped win the war. Addressing Congress one month after the armistice, he stirred the spirit of the nation with some of the most vivid language ever uttered by a president. Paraphrasing Shakespeare's Henry V, Wilson paid homage to the men who served and fell abroad. "For many a long day we shall think ourselves accurs'd we were not there, and hold our manhoods cheap while any speaks that fought with these at St. Mihiel or Thierrey. The memory of those days of triumphant battle will go with these fortunate men to their graves; and each will have his favorite memory. 'Old men forget; yet all shall be forgot, but he'll remember with advantages what feats he did that day!' "

In 1918, one year after American entry into the war, Wilson sensed an opportunity for peace. Issuing a program for "peace without victory," he proposed fourteen points to serve as a basis for ending the war. Calling for "open covenants of the peace, openly arrived at; freedom of the seas; removal of trade barriers, reduction of armaments . . . [and] equitable adjustment of colonial claims," he capped his proposals with a call for an international League of Nations to police the peace.

The Wilson proposal fell on receptive German ears as troops in the trenches between the Rhine and the Atlantic waited for an honorable end to their four-year nightmare. Here was a peace that the Germans could accept. It provided for no crippling reparations, no humiliating loss of sovereignty, no self-effacing admission of guilt.

Haggard and starving, the German people jumped at it. German socialists overthrew the monarchy and sent Kaiser Wilhelm II packing to Holland.

As Wilson personally led the American delegation to Paris in January 1919 to negotiate the peace treaty, he assumed a global moral authority that no other American president had ever enjoyed. It was evident to all of Europe that not only had Wilson won the war by bringing America in, but his fourteen points had ended the bitter and sanguinary conflict.

Heralded in capitals around the globe as the savior of freedom, Wilson seemed invulnerable as he toured the continent he had saved. Praised in extravagant terms, he had swiftly become the most popular political figure on the planet. A proclamation issued by French labor unions hailed his sagacity: "President Wilson is the statesman who has had the courage and insight to place rights above interests, who has sought to show humanity the road to a future with less of sorrow and carnage. Thus he has given voice to the deepest thoughts which stir the democracies and the working class." The *New York Times* reported from Rome that "the whole national life seems in suspense awaiting the arrival of President Wilson, to whom are attributed almost supernatural powers. All parties are one in this; to await the verdict of his judgment on the most widely differing subjects. No man from a far continent has ever had such a deep and powerful influence on the mind of the whole nation or has done so much to inspire them to high ideals and noble aims."

But Wilson, amid such idolatry, seemed to lose his political instincts—and much of his political capital. This man, who had survived and prospered politically by an almost infallible ability to read the thinking of his countryman, suddenly had his head turned by flattery. And both at home and abroad, people noticed.

According to the historian Elmer Bendiner, Harold Nicolson of the British delegation thought "the president seemed to suffer from the same delusion that had afflicted the French revolutionary leader Marat: that . . . the general will of the people was embodied in him alone. It was probable that Wilson cultivated that illusion long before he came to Europe, but there can be little doubt that the cheers of the millions and the florid acclamations of the press . . . were overpoweringly heady."

Woodrow Wilson had always appeared rather stiff-necked, humorless, self-righteous, and puritan. He didn't need mass idolatry to make him seem haughty. But the messianic treatment he received in Europe certainly did nothing to increase his humility. And his royal reception rankled Americans. Homespun senator Lawrence Y. Sherman told the Senate "I wish to compare the zinc garbage cans out of which the American soldier is fed with the . . . solid gold service and the inlaid mahogany table from which the president . . . [is] feasting in London." Condemning Wilson's reception as a "foolish display of un-American adulation abroad," the Senator attacked "the parasites who walk backward making obeisance to guests and other useless appendages of European pomp and circumstance."

Wilson had also alienated his political opponents by diverting his attentions to campaign actively for a Democratic Congress in the midterm elections. Unwisely, he "left his pedestal and entered into the bitter [Congressional] elections of November 1918, calling for a Democratic Congress that would help him win the war," historian Elmer Bendiner later wrote. When Republicans retaliated, Wilson's prestige seemed to be squandered on a political mission. Wilson's efforts failed; the Republicans, bitter at his intervention, took control of both houses of Congress.

As Wilson left American soil for Paris soon thereafter to negotiate the treaty, he arrogantly refused to invite any congressional Republican as part of the delegation, and included only a single Republican diplomat, Henry White, to represent the party that had just beaten him soundly at the polls. Theodore Roosevelt spoke harshly on the eve of Wilson's departure: "Mr. Wilson has no authority whatever to speak for the American people at this time . . . his utterances every which way have ceased to have any shadow of right to be accepted as expressive of the will of the American people."

To make matters worse, the treaty Wilson negotiated in Paris was terrible. The Fourteen Points were quickly forgotten as French leader Georges Clemenceau, bitter at the loss of his nation's youth and fearful of a resurgent Germany, demanded harsh reparations that would cripple the defeated adversary for decades to come. Even more galling was his demand that Germany accept the entire burden of guilt for having started the war in the first place—a finding that was neither fair nor helpful.

In the ensuing treaty, Germany lost "one-tenth of its iron foundries, a third of its blast furnaces, three-quarters of its iron ore and zinc," along with coal deposits in Upper Silesia; it lost control of its colonies and of the provinces Alsace and Lorraine; and the Rhineland was slated for fifteen years of French occupation. Furthermore, Germany was burdened to pay hefty war reparations—a requirement that would plunge it into a crushing financial depression.

The Germans were outraged, convinced that they had been duped by Wilson. With their troops still on French soil and no Allied troops on any German land, they had been induced to surrender by the Fourteen Points, only to be hammered by the Treaty of Versailles. Called upon to submit to the peace terms, German delegate Count Brockdorff-Rantzau protested vigorously that "in this conference, where we stand toward our adversaries alone and without any allies, we are not quite without protection. You yourselves have brought us an ally, namely the right which is guaranteed by the treaty, by the principles of peace. . . . The principles of President Wilson have thus become binding for both parties to the war, for you as well as for us."

The Germans were not alone in their anger. The Treaty of Versailles was the work of the old world order at its most rapacious. With the rest of the world asked only to sit and await their decision, three men—Clemenceau, Wilson, and Britain's David Lloyd-George—stood over a map in secret, determining the destiny of millions of people.

Despite the promises of the Fourteen Points, the treaty did nothing to assure freedom of the seas, and parceled out colonies and territories to the victors like so many pieces of candy. The Japanese were not accorded the statement of racial equality they sought, but got economic and political rights in northern China instead. Indeed, so thoroughly raped was China—one of the victorious powers—that she refused to sign the treaty.

The only shred of his idealism that Wilson managed to salvage was the League of Nations, established by the treaty. The League would be an assembly of the nations of the world committed to the maintenance of global peace. A precursor of the United Nations, the League would be empowered to ask its members to enact economic sanctions and to bring other pressures, to bear against aggressors. It was envisioned by Wilson as a primitive form of global government to enforce an international rule of law.

But the League was crippled from the start by the cynicism with which the treaty was received around the globe. Quite simply, observers of all stripes—Americans chief among them—felt manipulated, even conned, by the Versailles powers. As they watched, the old-order nations had reverted to their old balance-of-power games, and ended a war fought for idealistic principles with a treaty based on greed and vengeance.

And the League itself, embedded in this flawed treaty, satisfied nobody. Liberals and internationalists felt its mechanisms for enforcing world peace were too paltry to make much difference; conservatives and isolationists warned that the League would usurp national sovereignty.

The internationalist New York *Tribune* found the treaty fatally weak, complaining that it allowed for "no safeguarding of peace. To each nation is reserved liberty of action. There is no limitation of armaments, no international police force under the control of the League. . . . We have before us something in the nature of an Entente Cordiale such as was established between Great Britain and France when they agreed not to act together but to confer with a view to action when trouble threatened."

But to others, the League seemed to go too far. Isolationist senator William H. King, Democrat of Utah, worried that joining the organization would mean abdicating our "sovereign rights" by ceding power to commit the United States to international action from Congress to the League. Republican senator William Borah of Idaho was concerned that the treaty amounted to a "renunciation of the Monroe Doctrine" because it would grant the League, not the United States, the right and duty to protect the hemisphere from foreign colonization. Indeed, the New York *Sun* felt that the treaty establishing the League surrendered so many national prerogatives that a constitutional amendment would be necessary to allow the United States to sign the treaty.

Senator Henry Cabot Lodge, the Massachusetts Republican who headed the Foreign Affairs Committee with jurisdiction over the treaty, was particularly harsh in his condemnation, warning that it put European interests first. Lodge warned Wilson that "we are invited to move away from George Washington [who had warned, in his farewell address, against involvement in European power poli-

tics] toward the other end of the line at which there stands the sinister figure of Trotsky," a leader of the Communist Russian Revolution
then raging.

Chastened by the storm of criticism, Wilson returned to Paris to
amend the document creating the League to include an explicit
recognition of the Monroe Doctrine. Responding to further congressional demands, he included an escape clause that nations could
use to leave the League, and exempted internal disputes within a
nation from the League's purview.

But the visceral anger against the League only mounted in the
United States. Determined to fight for his brainchild, Wilson, though
now ailing, embarked on a backbreaking—some said suicidal—tour
through America to bypass Congress and enlist the support of the
American people.

"On the evening of September 3, [1919]," Bendiner writes, "Wilson, his face looking gray and old, distorted into spasms by [a] recurring nervous tick, plagued by almost incessant headaches," boarded
the presidential train. "The trip was to wind through the Midwest,
the Northwest, down the Pacific coast, and through the Southwest in
twenty-seven days, on a schedule calling for twenty-six major
addresses and scores of whistle stops."

Wilson dutifully took to platforms throughout the country to fight
for his League, but he had clearly lost his touch. He missed the point
entirely, framing his defense of the League in purely legalistic terms,
rather than tapping into the basic spirit of American patriotism and
idealism he had so effectively mobilized in winning the war. It is
impossible to read his speeches as he toured the nation without
being astonished at their pedantry and lack of uplift. Was this the
same man whose eloquence had inspired a nation to win the war and
Germany to give it up?

His speeches began with a lecture on land titles, making the driest
of cases for the ideal of self-determination of peoples. Wilson then
discoursed, at tedious length, on the histories of Poland, Upper Silesia, Manchuria, and a host of other areas equally remote and unin-
volving to his audience. By the time he got around to mentioning
that "the heart of the Covenant is that there shall be no war"—complete with a word-for-word recitation of the full text of Article X of
the treaty—his audience must have been lost in slumber.

On the defensive against the attacks on the League by Republican senators, Wilson devoted the balance of his speeches to a point-by-point rebuttal of the charges made by the League's enemies on issues such as whether it preempted U.S. sovereignty or failed to recognize the Monroe Doctrine. The total effect was numbingly boring, even insulting. As he ponderously reviewed each provision, he punctuated his remarks with the ultimate turn-off: "Every lawyer will follow me when I say . . ."

Having failed to invoke any patriotic reasons for ratification of the League, Wilson concluded by insisting that the nation must either accept the treaty as a whole or reject it altogether. He justified the demand by saying that "the rest of the world would find it very difficult to make any other kind of treaty and the world cannot breathe in the atmosphere of negotiations."

Wilson had clearly lost his feel for America; his waning health may have been partly to blame, but, by any measure his foundering campaign in defense of the League was nevertheless a political fiasco. Incredibly, Wilson even failed to invoke the memory of the more than 100,000 dead American soldiers who fell in the "war to end all war," though he believed, no doubt sincerely, that the treaty he was advocating would seal the newfound peace. He made no mention of American exceptionalism as he expounded the case for the treaty. Indeed, in Wilson's final address for the League, at Pueblo, Colorado, he mentions the names of other countries fifty-four times—twice as often as he speaks about his own nation.

Could it have happened another way? It's not hard to imagine a Ronald Reagan defending the League, linking it to the unique sense of altruism in American hearts. A de Gaulle might have argued for the treaty as a means to safeguard his nation's grandeur and traditions; a Churchill might have declared such an agreement yet another contribution of his old, historic nation to global civilization. But Wilson failed to present the League in anything like these terms.

After stumping through Missouri, Ohio, Kansas, Indiana, Nebraska, Montana, Wyoming, Washington State, California, Nevada, Utah, and Colorado, the president suffered a massive stroke on September 26 as he traveled from Colorado to Kansas. Returning to Washington, he remained essentially incommunicado for the duration of his term, watching helplessly from his sickbed in the

White House as the Treaty of Versailles fell under attack on the Senate floor. Henry Cabot Lodge proposed "fourteen reservations," a dig at Wilson's Fourteen Points. But the ailing president refused to entertain the argument further—and in the end failed to secure the two-thirds majority required for passage in the Senate.

Why did Wilson lose? Most historians have focused on Wilson's stubborn—some say senile—inflexibility in refusing to negotiate with Lodge. Convinced that he had already covered most of the issues Lodge was raising, he refused to return to Europe with his tail between his legs asking for more amendments.

In fact, Lodge's reservations were not really crippling; most of them are now embedded in the United Nations charter and procedures. According to Bendiner, Lodge wanted to stipulate that "the United States undertakes no obligation to interfere in any controversies between countries or to safeguard anything unless in each case Congress approves the contemplated action." The war powers amendment, passed in the 1970s in the wake of Vietnam, gives Congress essentially the same rights. He sought to vest the right to withdraw from the League in Congress, to decree that the United States could not accept any mandates to administer colonies without the consent of Congress, and to stipulate that domestic issues like immigration policy, labor laws, trade tariffs, and the like would be outside of the League's jurisdiction.

None of these reservations should have troubled Wilson. But even when Lodge indicated his willingness to settle for a whittled down list of reservations, Wilson wouldn't budge. In the end only fifty-five Senators backed the treaty, short of the required sixty-four; had Wilson accommodated Lodge's reservations, it would have passed easily.

Why Wilson wouldn't compromise is a mystery of history. But why he was so weakened that he had to is worth exploring. Why couldn't Wilson rally the nation to support the League? His physical infirmity is only part of the answer. For over a year, a healthy Wilson had the opportunity to make the case for the League of Nations to an American public that had voted for him in two elections and followed his lead into a victorious world war.

Of course, Wilson should have addressed the reservations of his Republican colleagues to get his treaty adopted; the flaws in the harsh and severe vision that emerged from Paris limited its viability.

Yet in the end his failure on the public stage was at least as important: Wilson's inability to rise to the occasion and convince Americans that the treaty was part of their nation's manifest destiny deprived him of the groundswell of public support that might have tipped the balance in his favor.

AL GORE RUNS AWAY FROM HIS ENVIRONMENTAL BELIEFS . . . AND LOSES AS A RESULT

Had Al Gore stood up for the ideals he really believed in, he—and not George W. Bush—would have been America's forty-third president. Gore lost power by neglecting his principles in the heat of the political campaign. In an age in which politicians win elections by inventing beliefs they have never really held, Gore stands out as an anomaly: A passionate ideologue who lost by dropping his signature issue—the environment—in what amounted to a failure of political nerve.

It's hard to look at a Churchill, Reagan, de Gaulle, or Lincoln and appreciate the self-doubts they must have suppressed as they faced adversity and electoral rejection by speaking up for their deeply held beliefs and core principles. But the tragedy of Al Gore offers an insight into how fierce the inner turmoil can be for one who tries to stand on his convictions even as they are ridiculed. These four towering men of history mastered their doubts. Gore's doubts mastered him.

To stand on principle, you have to run on principle. The results of the contested Florida election notwithstanding, there is a much more fundamental case to be made that Al Gore failed to become president because at a crucial moment he walked away from his defining political cause. Gore fought for the environment whenever he held public office; he wrote a best-selling book about it; it was the issue with which he was most universally identified. And yet he repeatedly buried the issue during his political campaigns. Figures like Reagan, Churchill, de Gaulle, and Lincoln ultimately achieved power by sticking to their guns; Woodrow Wilson and Barry Goldwater suffered defeat after pushing their points too hard.

In the 2000 election, Al Gore failed by not pushing hard enough.

To review how Al Gore came to drop the environment as his core issue is to learn how a candidate can be consumed by doubts, misled by advisors, and betrayed by caution. Reagan and de Gaulle never entertained serious self-doubts; Lincoln and Churchill overcame them. But in 2000 Gore's fears and worries eclipsed his capacity for resolution and action. And the resulting paralysis by analysis cost him the presidency.

It can be difficult to recognize a true believer in this age of image manipulation, of politicians who bite their lower lip to simulate emotion. But Al Gore was a true believer on the environment. His commitment ran as deep and as true as Reagan's, Churchill's, Lincoln's, or de Gaulle's. Even when polls said the public wasn't ready for a serious focus on the issue, he continued to study, ponder, and act when he could to save the planet from degradation at human hands.

The sincerity of his commitment makes his inability—even unwillingness—to speak out about his key issue when it counted even more tragic.

In the 1980s, Al Gore, the Democratic senator from Tennessee, was a man ahead of his time. Deeply concerned about phenomena such as global climate change and ozone depletion, Gore was an early voice signaling the onset of a new host of problems for the world's environment. As Americans gradually emerged from the cold war and the fear of nuclear devastation, Gore was there to highlight the degradation of the environment as the new threat facing our planet.

Gore often traced his environmentalism to a childhood spent on a family farm—an experience that taught him, he said, "my earliest lessons on environmental protection . . . about how nature works." Gore was influenced by his "mother's troubled response to Rachel Carson's classic book, *Silent Spring*," whose warnings about the perils of spraying the pesticide DDT launched the environmental movement in 1962. Later, in Vietnam as a reporter/soldier, Gore learned about the dangers of Agent Orange, a defoliant used in the war with tragic consequences.

In his best-selling book, *Earth in the Balance,* Gore recounted another formative force in his environmental journey. Roger Revelle, a Harvard professor, had dazzled the young Al Gore with his research on atmospheric carbon dioxide and global warming. As a congressman, Gore invited his old professor to testify before Congress. Where Gore expected his colleagues to be surprised, he recalled that it was, rather, his turn to be "shocked"—by their indifference.

Gore was way ahead of his time on the environment. A reading of his speeches in the 1980s and early 1990s, side by side with newspaper headlines a decade later, makes it obvious what a prophet he was. As early as 1987, Gore was writing about the environment; later, in his book, he wrote, "it is clear that doubling CO_2 (carbon dioxide) will, in fact, increase global temperatures and . . . subject us to . . . changes in global climate patterns." Warning that "civilization is now capable of destroying itself," Gore issued a sobering challenge: "we need to act now on the basis of what we know."

Gore was also becoming concerned that the growing ozone hole over the Antarctic, and the thinning of the protective layer elsewhere, would increase skin cancer rates worldwide. In an observation that may have then seemed arcane but sounds prescient today, he noted that "for every 1 percent decrease in ozone there is a corresponding 2 percent increase in the amount of ultraviolet radiation . . . and a 4 percent increase in skin cancer."

But "the real danger from global warming," Gore observed, is that "the whole global climate system is likely to be thrown out of whack." And his concerns didn't stop there: devoting equal attention to the prospect of extensive mercury pollution, oil spills, destruction of the rain forests, acid rain, DDT, overuse of antibiotics in livestock, and the depletion of global resources, in the 1980s Al Gore projected a vision of future environmental trauma that feels more real than ever in 2002.

As forthright in articulating solutions as he was in identifying problems, Gore proposed a "global Marshall plan" that would require a worldwide "cooperative effort," through voluntary treaties and accords, to establish "constraints" on industrialized society.

Gore realized that the responsibility for leading the way would fall to the United States. Yet he recognized that the political constituency was not there for dramatic action, and harbored concern that public opinion and scientific fact were running in opposite directions. Any effort to heal the environment, he foresaw, would have to take into account that "public attitudes are still changing," and he fully expected that "proposals which are today considered too bold . . . will soon be derided as woefully inadequate."

As he firmed up his national standing as an environmentalist, Gore sought to use his reputation to win the Democratic presidential nomination in 1988. Young and in a hurry, he felt that his national leadership on the environment might give him a shot at the presidency.

He was in for a rude awakening.

True to his principles, in announcing his candidacy on June 29, 1987, Al Gore proclaimed that one of the central reasons for his candidacy was "to try to elevate the importance" of the environment "as a political issue." Promising to confront threats to the environment, Gore said he would make environmental issues and global warming the "principal focus" of the campaign.

Instead of applause, he was greeted with ridicule. He looked back with pain at how the *New York Times,* which had not yet begun to treat environmental concerns with complete seriousness, derided his issues as "esoteric." Perhaps more predictably, conservative columnist George Will doubted the efficacy of Gore's environmental campaign, observing that "in the eyes of the electorate," the environment was "not even peripheral."

His Democratic rivals picked up the chorus of mockery. When Gore unfurled his signature issue at a Democratic presidential debate in 1988, he was met with a sour response. Contending that "this problem of the greenhouse effect is going to be one of the most severe environmental challenges we have ever faced in the entire history of humankind," he found his trial balloon was quickly deflated by fellow candidate Jesse Jackson, who observed acidly, "Senator Gore just showed you why he should be our national chemist."

Stung, Gore came to realize that his environmentalism wasn't selling. Two of his young consultants at the time—Jack Quinn and Bruce

Reed—wrote a fifteen-page memo advising the candidate to move on, jettisoning his pet ideas to focus on tried-and-true Democratic issues like the economy. Their advice: "the people are ready for a dose of responsible indignation on the question of economic fairness." Translation: get off the environment. But Quinn's and Reed's advice, and the ridicule of Jackson and Will, tapped into an inner seed of doubt already beginning to germinate in the insecure Gore. He began to wonder if his convictions made good politics. Like a cancer, the doubt first nagged, then grew, and finally ended up consuming his political effectiveness.

Gore caved in. "I began to doubt my own political judgment, so I began to ask the pollsters and professional politicians what they thought I ought to talk about," he recalled. "As a result, for much of the campaign I discussed what everybody else discussed . . . a familiar list of what the insiders agree are 'the issues.' "

Gore had come to realize that there was no profit in prophecy. "Not many people were focused on the global environment," he recalled. Most people thought only of the pollution in the air they breathed and the water they drank, rarely looking beyond to see the need for urgent measures to face global environmental challenges.

Determined to end the ridicule and mockery his farseeing vision had brought upon him, Gore stopped speaking about his key issue.

Off-balance, Gore seemed to hang onto his political viability through some southern primaries, but his candidacy sputtered and died when he was defeated by Massachusetts governor Mike Dukakis in the key northern states.

Imagine losing to Mike Dukakis!

The scar of 1988 lingered long in Al Gore's psyche. Even when the environmental issue eventually started to attract national attention, he grew timid about using it. Al Gore was much like Mark Twain's cat, who once sat down on a hot stove. Having learned her lesson, Twain observed, his pet would never again sit on a hot stove. But she'd never sit on a cold stove either.

In a self-critical passage in *Earth in the Balance,* Gore writes that he continued to look for ways to talk about the environment, but acknowledges that he "came to downplay it" in his speeches. When he called for action to reverse ozone depletion, he noted bleakly "not a single word was written in any newspaper in America about the speech or the issue." Gore admitted that most Americans didn't consider the environment an important issue—and that "I didn't do a good job of convincing them otherwise."

Gore felt discouraged and defeated. "The harder truth is that I simply lacked the strength to keep on talking about the environmental crisis constantly whether it was being reported in the press or not."

After the defeat in 1988, an odd pattern developed in Gore's political life. During election campaigns, he turned his attention away from the environment. Yet when he served in office he remained its dedicated advocate—the very opposite of standard practice for most politicians.

Gore's on again/off again approach to speaking out about the environment underwent another change after his defeat in 1988. Now, relieved of the burden of running for office, he became green again. Once more he traveled the world to track environmental damage, and in 1991 he began to write about global warming. He published his book, *Earth in the Balance,* the following year. Hailed as "an environmental Paul Revere" by the League of Conservation Voters, Gore was freed from his self-doubts.

Gore reported hopefully that his days of political trepidation on the environment were over. Instead, he wrote, "I have become very impatient with my own tendency to put a finger to the political winds and proceed cautiously. The voice of caution whispers persuasively in the ear of every politician, often with good reason." But Gore expressed confidence that he could rise above everyday political considerations and pursue his goals without fear. "When caution breeds timidity," he wrote, "a good politician listens to other voices. For me, the environmental crisis is the critical case in point."

Gore opted to sit out the presidential contest of 1992. But his life changed completely when candidate Bill Clinton tapped him as his running mate that summer. In tossing him the vice presidential nomination, however, Clinton may have advanced Gore's career—but crippled his new self-confidence in advocating environmentalism.

With this new campaign, of course, came new advisors. And even though Gore's strongest calling card was his environmental record, Clinton's advisors "muffled" Gore on his pet issue. According to Turque, they "were nervous about how ... [Gore's] mix of eco-spiritualism and big government prescriptives would be used as a weapon by the Republicans." While Gore did not run away from his position on the environment, "he carefully distanced himself from it." Once again, this pas de deux of off-year environmental advocacy and election-year retreat was repeated, with ever more devastating consequences for Gore's self-confidence.

As Pulitzer Prize–winning Gore biographer David Maraniss confirms, "Gore was often persuaded by the voice of caution, even on the environment, the issue that meant the most to him and for which he promised to take the most political risk. . . . There were good reasons for this . . . but those good reasons bred timidity."

The balance of power between candidates and their consultants is delicate. Advisors often have extensive experience with political races, and an objectivity that the candidate himself may lack. But it's the candidate's campaign. He or she cannot be a spectator at his own race for public office. He mustn't step aside, as Gore did habitually, and defer to those around him.

Gore's inability to stand up and insist on speaking out strongly on the environment may have been the product of a lifetime in politics shaped by the expectations of others. This son of a congressman who became a congressman, and son of a senator who became a senator, was now tapped by a new authority figure—Bill Clinton—to run for vice president. Unsure of his ground, insecure about his political instincts, he once again agreed to sublimate what he knew to be right to what others urged upon him.

As he ran for vice president—to the derision of Republican Bush, who mocked him as "ozone man," and Dan Quayle, who called his views "bizarre," Gore retreated deeper into his shell, moving ever further away from his own original thinking on the environment.

With their laser-beam focus on the economy, of course, Clinton and Gore triumphed in 1992—an outcome that could only have reinforced the new vice president's impression that the environment and election-year politics were like oil and water. Yet, to his credit, Gore focused on the subject once he was in office, and with a vengeance. Taking the National Security Council as his model, President-elect Clinton proposed to create a National Economic Council, to fight the recession; Gore answered with a plan for a National Environmental Council, coequal with the other two, to focus on ecological issues. Clinton rejected his proposal, but Gore had set the tone for his tenure as vice president.

Working firsthand with Al Gore in the Clinton White House, I was deeply impressed by the sincerity of his convictions about the environment. Down the hall from the Oval Office, Gore sat at his vice presidential desk, under a gigantic NASA picture of the beautiful blue Earth as seen from outer space.

When the Gingrich-led Republican Congress slashed the federal budget and closed down the government after Clinton wouldn't agree to their cuts, the president's strategy called for focusing on the Medicare, Medicaid, and education cuts in the GOP agenda. Clinton would have downplayed the environmental cuts, because they didn't poll well as a major electoral concern. Disregarding the data, though, Gore insisted that Clinton add the environment to the litany. Whenever he screened a proposed advertisement for the budget war against Gingrich, he refused to approve it unless the environment were added.

The real Al Gore was back!

Gore's environmental record was the most activist of any vice president in history. Yet it wasn't enough to satisfy many true believers. Accusing Gore of errors of omission, environmental activists piquantly branded this supposedly passive period "Al Gore's Silent Spring." In 1997 the *New York Times* highlighted the dispute, writing that Gore, "one of the environmental movement's steadiest allies, is under attack from major conservation groups in an unlikely turn that carries important implications for environmental policy as well as presidential politics."

But the critics ignored that Gore was forced to battle for the attention of a president far more focused on economic development than environmental protection. Despite this, Gore worked actively to promote ecological issues throughout his eight years in office. Observers of his tenure as vice president naturally assumed he would make the environment a centerpiece of his campaign for the presidency in 2000.

But they weren't thinking of this candidate's history. Al Gore, after all, had taken away one lesson from his history of campaigns for national office: When election year rolls around, it is safer to focus on what the people are worried about than to dwell on your own pet projects—no matter how pressing those projects may be.

As Gore mounted his campaign, his memories of how environmentalism had failed him in 1988 dominated his thinking. Facing an unexpectedly strong challenge for the nomination from former New Jersey senator Bill Bradley, Gore back-burnered the environment entirely. In their first candidates' debate in January 2000, Gore devoted only two sentences to the issue: "I believe very strongly in protecting the environment. And I know we can do it in a way that protects our way of life and standard of living."

In each of the subsequent debates with Bradley, Gore repeated the pattern. Focusing heavily on the issue of patient protection from HMOs and the need to provide health care for the uninsured, Gore thoroughly neglected the environmental issues that had been so central to his career. By February, it was all Gore could do to offer a simplistic mantra—"I think that we ought to have clean air and clean water"—that belied his own sophisticated grasp of the issue.

Gore's refusal to speak up on the environment puzzled many, and left his supporters aghast. As early as April 1999, a writer for *Time* wondered if Al Gore had lost faith in his environmental principles. "Now that Gore is running for the White House and preparing to step out of Clinton's shadow, environmentalists and other voters want to know how green the Vice President really is. Does his record on the environment as Clinton's right hand man match the exalted and ambitious rhetoric of his book or has he, as he phrased it in *Earth in the Balance,* succumbed to the 'tendency to put a finger to the political winds and proceed cautiously'? In other words, is Al Gore the candidate the same guy who wrote the book?"

Time urged Gore to step forward and "stir up a tempest about climate change," but noted ruefully that "so far, it appears unlikely that Gore will do so. His strategists figure, quite rightly, that he can't be elected President solely as Mr. Environment and Technology. . . . If all this means that Gore will soft pedal his signature cause, climate change . . . that's bad for the earth and unworthy of a politician who has a record for being principled and decisive." The magazine closed by wondering, along with millions of Americans, "who is Al Gore if not the country's leading environmentalist?"

Who *was* Al Gore? As he entered the 2000 presidential battle, he was no longer the environmental warrior he had been as he prepared to run in 1988. That Al Gore was gone, buried under a decade of doubts and advice to chill his rhetoric on global warming. Memories of how Bush's father had mocked him as "ozone man" in 1992 must have haunted him. Insecurity and fear had won the battle for Al Gore's mind.

Gore's failure to address the environment became even more obvious during his three televised presidential debates against George W. Bush. In the entire series of debates—comprising four and a half hours of discussion—the environment was discussed, by the two candidates combined, for less than fifteen minutes. In two debates, it was mentioned only perfunctorily.

In the debates Gore seemed to return obsessively to economic issues, regardless of the subject under discussion. He dwelled on the need to protect Social Security tax revenues, mentioning the word *lockbox* seven times, and alluded to his charge that the Bush tax cut was targeted at the "wealthiest one percent" even more frequently. The environment was discussed only in the second presidential debate, on October 11, 2000.

It became evident to voters watching the Bush–Gore debates that the Democratic candidate was simply not sure who he was. Not only was he running away from his fundamental principles, but he seemed unable to find his own persona. In the first debate, Gore emerged like a political Mike Tyson—raging, punching, even biting, furious that Bush was horning in on his Social Security, health care, and Medicare issues. Chastened by a harsh public backlash, a sweet, generous Gore came out for the bell in debate two, sharing credit magnanimously with his Republican rival and pointing up their differences in ever-so-gentle tones. By the third debate, Gore seemed as confused about who he was as the voters were.

Even as Gore was sublimating his environmental views, he faced a surprising new challenge on the left—one that ultimately proved fatal to his candidacy. Activist Ralph Nader, running as the nominee of the Green Party, made it part of his campaign to discredit Gore's reputation as the most prominent "green" in America. Doubtless aware that the attack could make some serious inroads among liberal

Democrats, Nader began accusing Gore of selling out the environment.

On November 3, with the election only hours away, the *New York Times* concluded that Nader had "succeeded in driving home the notion that Mr. Gore's true colors are far less green than he has made out." Gore must have been livid to hear himself flanked on the left by Nader on the environment. The vice president had always regarded himself as untouchable on environmental issues. Now Nader was ridiculing him for inconsistency on this, of all issues.

Indeed, in the closing weeks Gore had proven unable—or unwilling—to gain traction on the environmental issue. The *Times* noted that it had been difficult "for Mr. Gore to delineate the deep differences over the environment between himself and Mr. Bush." This was truly amazing. Al Gore, who had boldly staked out the environmental turf fifteen years earlier, had gained *no* advantage over Bush on the issue. It was as if Richard Nixon had received no credit for a tough stand on law-and-order, or Reagan was bested on the issue of tax cuts.

How should Gore have handled Nader? The way Bush handled Pat Buchanan.

Bush, too, had his distraction—Buchanan, who was running as a third (or was it fourth?) party candidate on a hard-right platform. But where Gore would not move to the left on the environment to cut off Nader's challenge, Bush stood on his principles admirably and stole Buchanan's themes by proposing a huge trillion-dollar tax cut. With Bush pushing for a sharp, steep slash in tax rates, Buchanan's right-wing insurgency could gain no traction.

After the election, Buchanan admitted to me, "I spent election night praying that I wouldn't take enough votes from Bush to elect Gore." Nader made no such prayers.

Of course, the election of 2000 would be like no other. As became clear late in the evening after the polls closed, the contest came down to who would carry Florida. And here is the ultimate irony: If any state in America is environmentalist, it's Florida.

Residents of the Sunshine State had long been attuned to man's perilous relationship with nature; concerns about offshore oil drilling, protecting the Everglades, and preserving the state's famous beaches had been a staple of local debate for years. But, during his campaign stops in Florida, Gore had spent precious little time discussing the environment. Even after Nader had begun to hit him hard Gore had stayed away from the issue, speaking only three sentences about the environment during a stop in Tampa a few days before the election.

In the end, Ralph Nader drained away ninety thousand liberal votes from Al Gore in Florida. And that made all the difference.

Why did Gore run away from the environment, even after his connection with his core constituents was imperiled by the attacks of the Green Party?

When Gore should have stood on principle, why did he flinch and falter?

One key Gore advisor told me the candidate was afraid that the more radical parts of his book *Earth in the Balance* would come back to haunt his campaign. "We would go into meetings with Gore urging him to speak out about the environment, but he would always make clear that he was afraid of being accused of supporting higher taxes as a result of what he had advocated in his book."

Another Gore campaign aide confirms this image of a candidate preoccupied by fears of attack. "He didn't talk about tobacco because he was worried about being called a tobacco farmer," he told me. "He wouldn't discuss Internet taxation because he didn't want to be reminded that he had claimed to be the father of the Internet. And he wouldn't talk too much about the environment because he was afraid his book would be used against him."

A pessimist at heart, Gore was as painfully conscious of the dangers of advocacy as Clinton was aware of its possibilities for self-advancement. To Bill Clinton, born poor and without social connections, politics was full of opportunities to win votes. To Al Gore, heir apparent to his father's political career, each such opportunity was rife with pitfalls that might derail his well-planned march to the top.

Needless to say, Al Gore would have been ideally suited to tap into this latent commitment to the environment, particularly among younger voters. Had he only demonstrated the courage of his convictions, he could have blunted the appeal of Nader and the Green Party, won the hearts of the environmentally conscious voters of Florida, captured the presidency, and carried on with the activist agenda to which he has always been committed.

Instead Gore was sidetracked by fear, mounting a lackluster bid for the presidency that relied more on his old boss's preoccupations—the economy, Social Security, and so forth—than on the issue for which he had proved his passion. Even after the polls had closed and the counting and court battles began, Gore outsmarted himself once more. Instead of taking a stand on principle, he adopted an opportunistic position—to his own detriment.

When the slender margin of Bush's lead in Florida became clear, Gore faced a choice between taking a principled position, demanding that all the votes in Florida be recounted, or restricting his request to a select number of counties. Initially he called for a statewide count, but quickly backtracked on legal advice, focusing on only three counties. His lawyers had persuaded him that these three areas were most likely to augment his vote, and that no one could predict what a statewide recount would show.

By sacrificing principle and demanding a statewide recount, Gore lost the moral upper hand; now it looked like he was trying to introduce legal wrangling and procedural infighting to the contest. As it turned out, poor Al Gore was once again too cute by half.

The *New York Times* subsequently showed that had Gore's demands been met, and the three counties in question been recounted, he would still have lost the state to Bush. But, the newspaper noted, if the entire state had been recounted—the principled position Gore took initially and then abandoned—he would likely have won.

Politicians are people. Subject to insecurities, they sometimes rely too much on experts. Experts, for their part, too often cling to the safe advice, and hew too closely to the conventional wisdom.

In the election of 2000 focusing on the environment was seen as risky, while talking about Social Security and health care seemed safe. Afterward, demanding a recount in three counties appeared prudent, while asking for a statewide count was chancy.

But in both cases the risky course was the right one, and the safe one would bring about defeat. Whether the brass ring will come around for him again is unclear, but in the unforgettable contest of 2000, Al Gore lost because he failed to stand on principle.

STRATEGY TWO

TRIANGULATE

There is nothing less rewarding than political stalemate, in which two sides push with diminishing force, year after year, to resolve intransigent political issues. As in World War I, when millions of entrenched Allied soldiers faced millions of equally stalwart Germans across a no-man's-land that stretched from the English Channel to the German border, such standoffs beg for a flanking movement, a new initiative, as an alternative to the drudgery of attrition.

In politics or any other vocation, static, unchanging conflicts deaden the imagination, dampen the spirit, and sap the capacity to achieve eventual agreement and consensus. Rather than continue endless quarrels or debates—in the workplace, the marketplace, or the polling place—there are times when the best solution is to rise above the confrontation, embrace the best solutions from each side, and formulate a third way to deal with our problems.

Over years of mind-numbing debate, particular ideas, solutions, and proposals tend to become identified with a single party or faction—even when the problems they address are common to both. Democrats may want gun control, while Republicans suggest longer sentences—but crime is the common enemy of both parties. Democrats may want more education spending, while Republicans embrace vouchers—but improving schools is their common goal.

With the passage of time, each of the two American political parties has come to be identified with certain frustrations and concerns.

In the modern era, crime, taxes, defense, government regulation, immigration, morals, government spending, and deficits have all been Republican issues. Education, the elderly, health care, choice, poverty, child care, homelessness, the environment, and racism have been Democratic concerns.

Taken to extremes, of course, this linking of parties to problems is ridiculous. After all, neither side is pro-pollution, or pro-hunger, or pro-unemployment, or pro-poverty.

But somehow we have come to associate not only certain solutions, but also the underlying problems, with one party or the other. We are truly surprised when a Democrat speaks about crime, or a Republican shows compassion for the poor. We wonder if they are deliberately blurring party lines, or trying to sell us a bill of goods. But it should not be so. Both parties recognize the gravity of all of these concerns; they merely differ in their priorities, and in the solutions they believe will be effective.

The identification of certain problems with certain parties or factions opens up a magnificent strategic opportunity: the chance to solve the other side's problems. When one resolves the issues normally identified with a rival, one removes its fundamental raison d'être and diminishes its chances of future success. The better education, the environment, and the elderly are, the less motivation there is to vote Democratic. The lower taxes, crime, and welfare get, the fewer reasons there are to elect Republicans.

In the workplace or other settings where politics abounds, the same strategy can be decisive. Solve the problems that keep the other side in business, and it will go broke. Give them what they want and they will go away.

In the Clinton administration, we called it triangulation.

The idea behind triangulation is to work hard to solve the problems that motivate the other party's voters, so as to defang them politically. If you are a Democrat, balance the budget, reform welfare, cut crime, and watch voters flee the GOP. If you are a Republican, improve education, lower poverty, and watch your ranks swell.

But in solving the other side's problems, no politician can afford to become a carbon copy of his opponent, saying "me too" when his adversary focuses on an issue and "I agree" when he proposes a solution. The essence of triangulation is to use *your* party's solutions to solve the other side's problems.

Use your tools to fix their car.

The triangle in question is equilateral, with the apex suspended over the middle of the base. Triangulation involves using the solutions of both parties to solve each new problem. It involves adopting what is best from each party and formulating a third approach that discards the failed solutions and embraces those that work.

Cut welfare—not just by establishing time limits and work requirements, but by providing job training, child care, and tax credits to create jobs.

Hold down crime—not only by hiring more police or building prisons, but by limiting handguns and banning assault weapons.

Improve education—by increasing spending, but also by using vouchers to encourage competition between private and public schools.

Protect the environment—in part by regulation, but also by creating a market in pollution permits where industries pay one another to cut emissions.

But in combining the best in each party's agenda, the successful triangulator leaves behind the worst. While Clinton adopted certain Republican ideas on welfare reform, he sturdily refused to allow any cut in food stamps or child nutrition for families on welfare. (He did allow, as a tactical concession, cuts in such aid to legal immigrants, but then reversed these reductions the following year.) Bush listened when Democrats advocated increased aid to urban schools, but insisted on formulating alternatives for failing schools.

This synthesis of the best in each party's agenda amounts to more than a bisection, a splitting of the ideological difference. Triangulation does not call for a line split in half. It traces a path in which the best of each party comes together in a place higher and better than either would reach on its own.

Of course, solving the problems that animate the other side's agenda is only half the battle. One must also push a strong program to solve the problems of one's own party. It is fine for a Democrat to cut crime and slash welfare rolls, but he must also take care to improve schools and clean the environment. A Republican can urge more focus on education, but he'd better push tax cuts at the same time.

These, then, are the three principles of triangulation:

1. Solve the other side's problems.
2. Use solutions from both parties to do so.
3. Continue to focus on your own issue agenda.

In *The Tao of Power*, Lao-tzu writes that "to hold to the center is to listen to the voice of the inner mind." In their "inner minds," Americans come to certain conclusions as they witness the swirling political debate around them. Triangulation, as a methodology, involves a move to the center. As an ideology, it calls for turning down the volume of one's own ideas and listening, instead, to the voices of the "inner minds" of one's nation.

When Bill Clinton moved his party from a welfare state mentality to a focus on the tradeoff between opportunity and responsibility—we'll give you welfare, but you must work—and when George Bush moved his from a focus on small government to one on compassion, both were listening to the "inner minds" of their nation.

In each of their personal lives, baby boomers had come to see the failures of the limitless hedonism of our youth, and the straitjacketed self-denial of our parents. The result was that their "inner minds" spoke of the need for opportunity to replace self-denial, and for responsibility to take the place of hedonism.

As Generation X saw the difficulty of getting and keeping jobs in the global economy, they began to worry about the education their children were getting. Their "inner minds" called for tougher schools and stricter standards, a mantra that came to dominate George W. Bush's political agenda.

In the chapters that follow, we will examine how America's two most recent presidents have been elected largely through a strategy of triangulation. By crossing over and addressing the core values of the other party, both George W. Bush and Bill Clinton were able to expand their base and win their elections.

From the outset of his campaign, George W. Bush was determined to take education and poverty away from the Democratic Party and make them, in part at least, Republican issues. Even as he announced his candidacy, he declared it his mission to "match a conservative mind with a compassionate heart." He refused to concede issues like education, poverty, and health care to the Democrats, but determined to make them mainstays of his campaign. How did he succeed? How did the Democrats let him co-opt their best issues? And why didn't doctrinaire Republicans knife him the back while he was trying?

Whereas Bush felt it was time to reverse decades of Republican insensitivity to the problems of the poor, Bill Clinton ran for office demanding that his own party address the difficulties of the middle class. To the "new Democrat" Bill Clinton, crime and welfare were not hot-button right-wing code words for racism, but the just demands and grievances of the middle class. Like Bush, he wouldn't concede this turf to his opponents. Pledging to "end welfare as we know it" and backing capital punishment and the balanced budget, he declared that "the era of big government is over"; repositioning himself and his party, he reversed a twelve-year losing streak in national elections. How did he manage to bring along his party, kicking and screaming, as he led it to a new national agenda?

But the most dramatic example of successful triangulation comes not from America but from France. In 1981 her new Socialist president, François Mitterrand, nationalized vast segments of the French economy. But with Reagan and Thatcher tacking right as he moved left, France's capital drained across the Channel and the ocean and the nation fell into a deep recession. Powered by France's difficulties, rightist candidate Jacques Chirac won the legislative elections of 1985. In response, Mitterrand triangulated. He let Chirac privatize all that he had taken over four years earlier. Then, deprived of an issue, Chirac lost his 1987 presidential bid against Mitterrand. Triangulation had swept away Chirac's chances of victory. His own success dismantled his electability. How did Mitterrand keep his party in line as he reversed course? How did he pull it off?

And yet not everyone who triangulates wins. Nelson Rockefeller, the liberal Republican governor of New York, found himself falling between two parties when he tried to forsake the orthodox GOP agenda and embrace civil rights and social legislation more characteristic of Democrats. Why did he fail to carry his party along with him as successfully as Bush, Clinton, and Mitterrand? How did his attempts at triangulation come to seem phony to Democrats and traitorous to Republicans?

GEORGE W. BUSH MOVES THE GOP TOWARD COMPASSIONATE CONSERVATISM

As he pondered his race for the presidency, George W. Bush realized that the Republican Party could not run another campaign using the same formula that Reagan and his father had used. It wasn't working anymore, as the Republican defeats of 1992 and 1996 had made painfully clear.

He had to prove he was not his father's Oldsmobile.

The Reagan–Bush formula had stressed a strong opposition to communism abroad, and a focus on less government and lower taxes at home. Socially, Reagan and Bush had also exploited latent middle-class resentments against crime and welfare, with George H. W. Bush memorably inveighing against the release of rapist Willie Horton by his Democratic opponent, Governor Michael Dukakis of Massachusetts. Culturally, Reagan and Bush hewed to pro-life, pro-gun positions to solidify the support of the Christian right. It was an approach that had won the GOP three consecutive national elections from 1980 to 1988.

But now things were different. It wasn't that voters had changed their minds; it was that the issues had just gone away. There was no more international communist menace. The budget was balanced. The deficits had disappeared (only to reappear again the future, like a magician's rabbit). Crime was down. Welfare reform was working. Lower unemployment had reduced fears about immigration.

The situation Bush and the GOP faced in 2000 was akin to that which confronts many companies, organizations, schools, or civic groups each year. Over time, every such group develops a certain image and its branding creates expectations that become indelible.

But when the things your company or group is known for are no longer necessary, what is left? How can you cope with the consequences of your own success?

The Republican Party faced just such a conundrum in 2000. It was out of issues.

Bob Dole, in mounting his ill-fated 1996 presidential race, had tried to squeeze one last victory out of the traditional Republican agenda. Criticizing Clinton as a "closet liberal," the Republican nominee called for the elimination of the federal Department of Education and demanded ever tougher measures against crime, immigration, and affirmative action. He campaigned on the charge that Clinton would raise taxes in his second term and doom the country to rising deficits by increasing federal spending.

But nobody bought it. Dole's warnings made little sense to a nation that watched as the budget deficit dropped, welfare rolls declined, and crime decreased. Turning aside Dole's forebodings, Americans reelected Clinton by nine points, giving him nearly 50 percent of their votes even though H. Ross Perot had made the contest a three-way affair.

Reflecting on Dole's self-destruction, George W. Bush saw clearly that he could not win on the old issues. He had to rebuild the Republican agenda and move his party to the center, much as Clinton had done for the Democrats. So Bush decided to return Clinton's favor. If the Democrats could cross over and solve his party's problems, he could do the same to the Democrats and make health care, education, and poverty into a new Republican platform.

It was a transition that seems to have come naturally to Bush, who never evinced much comfort with his own party's traditional rhetoric. Nowhere in the Republican posturing about morality, protection of life, deregulation, tax reduction, or individual liberty did he find the emphasis on compassion, love, or care for the needy that colored his own worldview.

In *A Charge to Keep,* a campaign autobiography created to introduce this southern governor to a nation unfamiliar with his ideas, Bush seemed eager to expand on the "kinder, gentler" imagery his father had dabbled in during his presidency. "A conservative philosophy has sometimes been mistakenly portrayed as mean spirited," the book observed. "Some who would agree with a conservative philosophy have been turned off by a strident tone." It was an echo of Bill Clinton's own call for a "government that is leaner, not meaner" in his 1992 speech at the Democratic convention.

Bush, too, based his campaign on setting a different tone. He determined to "prove that someone who is conservative and compassionate can win without sacrificing principle." Once again reminding one of Clintonian rhetoric, Bush observed that "an old era of American politics is ending," and promised to show that "politics, after a time of tarnished ideals, can be higher and better." As the country moved on from the Lewinsky scandal, Bush said he wanted to give America "a fresh start after a season of cynicism."

Embracing the language of triangulation, Bush's philosophy was that the work of his war on poverty would be "done by churches and synagogues and mosques and charities," which he saw as "a quiet river of goodness and kindness that cuts through stone. Some call their efforts crumbs of compassion; I say they are the greatness of America."

Criticized by many in the GOP for moving to the left, even as he went to Iowa to announce his presidential candidacy, Bush used the podium to lash back at his critics. He worried that his party's almost reflexive opposition to Democratic spending programs threatened, in the words of the poet William Butler Yeats, "to make a stone of the heart."

"I know this approach has been criticized," he said that June. "But why? Is compassion beneath us? Is mercy below us? Should our party be led by someone who boasts of a hard heart? I know Republicans—across the country—are generous of heart. I am confident the American people view compassion as a noble calling. The calling of a nation where the strong are just and the weak are valued."

The much-discussed key to Bush's campaign was his claim to the title "compassionate conservative." "I am proud to be a compassionate conservative," he said in Cedar Rapids in June 1999. "I welcome the label. And on this ground I'll take my stand." Bending and shaping his party's traditional policies to stress his emphasis on compassion, Bush, again almost paraphrasing Clinton, said in his autobiography that "I worked to reform welfare because I believe it is far more compassionate to help individuals become independent than to trap them in a cycle of dependency and despair."

From the moment Bush announced his candidacy, and until his inaugural speech, he kept moving his party to the center, reaching voters who had been turned off by the hard-right agenda of conservatives like former House Speaker Newt Gingrich. Launching his candidacy, Bush demanded that his party "match a conservative mind with a compassionate heart." Commencing his presidency, the forty-third president told Americans that theirs was a land of inclusion and equal opportunity, where "everyone belongs and everyone deserves a chance." Throughout his inaugural address, Bush steered clear of such core GOP issues as abortion, welfare reform, and opposition to affirmative action.

Bush's new twist on traditional Republican ideology was evident in his treatment of nearly every issue he addressed during the campaign. "It is conservative to reform welfare by insisting on work," Bush declared. "It is compassionate to take the side of charities and churches . . . It is conservative to confront illegitimacy. It is compassionate to offer practical help to women and children in crisis."

What lay behind Bush's inclusive strategy? Demographics, for one thing. Bush realized that the Republicans could no longer win with only white votes.

The GOP was running out of whites.

Determined to win minority votes, Bush was eager to reverse the racial dynamics of the partisan divide in America. In the past, Republicans had treated the black and Hispanic vote as a golf handicap. In their calculus, GOP strategists would automatically discount the 12 percent of America that was black and much of the 12 percent that was Hispanic and try to win the election by carrying three-quarters of the white vote.

But the old GOP math was increasingly becoming untenable as the Hispanic population swelled. Between 1990 and 2000, census statistics revealed that the Hispanic-American population rose by 60 percent. Suddenly the number of Hispanics equaled the black popu-

lation. If trends continue, in another decade Hispanics will vastly out-
number blacks, and whites will make up a smaller and smaller proportion of the vote.

Central to the Bush revision of the GOP message was a demand
that the party expand its base and appeal to minority voters previously antagonized by Republican positions on issues dear to them.
Announcing that "we are now the party of . . . idealism and inclusion," Bush pushed the party to drop its opposition to legalizing the
status of Mexican-Americans in the United States. He refused to follow the English-only focus of much of the party, and rejected calls for
an end to education benefits for illegal immigrants. Like a corporation reaching out to target new markets, Bush's campaign pursued
Hispanics, well aware of how effectively their political conversion
could swell the ranks of GOP voters in the twenty-first century.

As he campaigned, Bush seemed like a hybrid of the traditional
Republican conservatism of his upbringing and the compassion and
spiritual values that seem to have flowed from his later battle with the
bottle. In the pages of *A Charge to Keep,* Bush describes his fortieth
birthday celebration at the Broadmoor Hotel in Colorado, which led
to his decision to quit drinking. "The seeds of my decision had been
planted the year before, by the Reverend Billy Graham," he recalled,
who "planted a mustard seed in my soul, a seed that grew over the
next year. He led me to the path, and I began walking. And it was the
beginning of a change in my life."

Seeking to translate his new spirituality into political action, Bush
came repeatedly to emphasize the need for inclusion, "tolerance and
respect for the religious views of others." Calling himself "a uniter,
not a divider," he spoke of developing a "morally grounded and
socially inclusive" political agenda that departed from the traditional
focus of the right and tapped into "a much greater tradition, a tradition of social justice." He came to feel deeply that "our prosperous
society must offer answers for the poor," an item not traditionally
high on the list of conservative concerns.

And though many Democrats might have believed otherwise, this
compassionate crusade was not merely opportunistic. George Bush,
Sr., may have experienced a road-to-Damascus conversion to conservatism when he became Reagan's running mate in 1980, but before
that the Bush family had been centrist, an offspring of the moderate

Republicanism of Dwight D. Eisenhower. And surely Bush's Clintonian outreach to African Americans must have surprised traditional Republicans. In his memoir, Bush describes how he was "shocked" by the assassination of Reverend Martin Luther King, Jr., and "horrified" as he watched snarling dogs and billy clubs "directed at America's own citizens" during the civil rights movement in the South: common reactions, perhaps, but notable for their mention in a southern Republican's autobiography.

Running for president, Bush candidly told a July 2000 NAACP convention that "the history of the Republican Party and the NAACP has not been one of regular partnership"—a confession his audience may have thought the understatement of the year. Bush spoke of "racial harmony" and "economic opportunity" as well as the need to "acknowledge our past." It was a subject that seemed to bring out a kind of domestic statesmanship in Bush: "For my party," he lamented, "there's no escaping the reality that the Party of Lincoln has not always carried the mantle of Lincoln." But he held out an olive branch, and expressed what seemed a sincere hope of finding a common ground.

This move toward inclusiveness would culminate in the surprising— some would say dubious—vision of multiculturalism that Americans witnessed at the Republican National Convention. To demonstrate the change in party attitudes, Bush asked African American general Colin Powell, his future secretary of state, and Condoleezza Rice, his future national security advisor, to deliver televised prime-time speeches at the convention. But the more memorable spectacle was the appearance of a black gospel choir, which belted out rhythms to the gathering of diversity-challenged delegates on the floor. A casual TV viewer, channel surfing, could be forgiven for thinking that he had switched to the wrong channel. Was this really the *Republican* convention?

Though critics called the gesture heavy-handed, and pundits heaped ridicule upon it, Bush felt it was in line with his efforts to attract African Americans and Hispanics who had traditionally been beyond the reach of his party.

Bush's pursuit of the African American vote was to prove unrequited—they overwhelmingly backed Al Gore. Yet Bush did much better in his outreach to Hispanics, no doubt because of his Texan background. In Arizona, during the first Republican debate, Bush addressed Hispanic-Americans directly: "Let me start by this way: *Muchos espanos viver en ese estado*—There's a lot of Hispanic Americans who live in this state and a lot who live in my state as well, which is a reminder that our party must broaden our base. I've tried to use my compassionate conservative message to do just that in the state of Texas, and all across the country. How do you intend—how do you intend to reach the growing number of Hispanics? And how do you intend to attract them to the Republican Party?" Even Utah's strait-laced Orin Hatch, a rival for the nomination, was moved to concede Bush's advantage: "I think that's one of the strongest things you have going for you." On the race issue, no other GOP candidate was equipped to match the Texas governor's record.

A candidate seeking to move his party to the center usually has to have a signature issue to demonstrate to a wary electorate that he means what he says. With Clinton, it was his promise in the 1992 campaign to "end welfare as we know it." With Britain's Tony Blair it was his commitment to curbing the power of labor unions. For George W. Bush, that signature issue was education.

The Republican Party had long been weak on education. But George W. Bush changed the tune, calling it "my top priority" in his first State of the Union address. It was an emphasis that dated back to the very start of his candidacy. Rejecting the GOP dogma that education was only a local concern, Bush asserted in his announcement speech that "it is conservative to insist on education standards, basics, and local control. It is compassionate to make sure that not one single child gets left behind."

In his focus on schools, Bush tapped into a core Clinton demographic: the soccer moms whose votes determined the Democrats' 1996 success. While Americans had always been concerned about education, before Clinton became president they tended to see it as a state and local issue. Schools may have dominated politics at the state capital, but they played little role in Washington thinking. Only with Clinton's concerted focus on increasing federal education spending did Americans come to see education as a national responsibility.

In 2000, George W. Bush decided he would take the education issue away from the Democrats—both to win the votes of soccer moms, and to convince voters of all stripes that he was moving his party to the center. Campaigning to be the "education president," he called for higher education standards and proposed a national program of academic testing to evaluate schools, targeting special aid at those whose scores lagged.

As the delegates gathered for the Republican National Convention of 2000, they may have been surprised to read that one of its four nights would be devoted to education. As late as 1996, the Republicans had insisted the issue had no place on a federal agenda. But that night, one after another, GOP orators spoke of the cardinal importance of reforming America's schools. Along with "compassionate conservatism," perhaps the single most memorable phrase of the Bush campaign was "leave no child behind," and it was a message the candidate hammered home relentlessly. In the coup de grâce, Laura Bush—the un-Hillary—emerged to describe her experience as a teacher and her husband's deep commitment to schools and children. As late as the third presidential debate against Al Gore, Bush was still attacking the lowered expectations for minority students implicit in lower standards for poor schools. He pledged to close the achievement gap, saying "There has to be a consequence [for low scores] instead of [just] the soft bigotry of low expectations."

Where Clinton used the tools of the left (gun control, job training) to solve the problems of the right (crime, welfare), Bush was reversing the formulation, using the tools of the right (vouchers, faith-based initiatives) to address the problems of the left (education, poverty). Calling for vouchers to give children of low-income parents a choice of schools, Bush pounded away at the theme in the debates with Gore, insisting that vouchers could help schools. Vouchers were a traditional GOP plan, but Bush put a spin on the ball that practically made him sound like a liberal. "When we find children trapped in schools that will not change and will not teach . . . there has to be a consequence. And the consequence is that the federal portion of federal money will go to the parent. . . ." By stressing how money would go *to* Mom and not *away* from the public school, Bush made the voucher system sound almost like a traditional Democratic welfare program.

During his first year in office, Bush continued his focus on education, according it the largest increase in his first budget. Pushing through a comprehensive program of testing, evaluation, and aid for troubled schools, the Republican president succeeded in capturing the issue for his party. So successful was his focus that a Fox News/Opinion Dynamics poll in July 2001 showed that more people trusted the Republican Party than the Democrats to deal with education. The Democratic issues had long been the three Es—education, environment, and elderly. Now there were only two.

Bush's very public embrace of education as a signature issue was an instructive study in triangulation. Like any attempt to triangulate, whether in politics or in business, it involved a symbolic outreach, a step that crosses the line and announces a new vision—a new brand, as it were—to a skeptical public. Issues, in politics, afford this opportunity. In other fields of life, the introduction of new product lines or services, new retailing strategies, and fresh advertising serve the same purpose.

Bush did the same with his other signature mission—the alleviation of poverty. Here was an issue the Democrats had owned since the Great Depression, yet their traditional solutions no longer held sway after Clinton enacted welfare reform. Bush wanted to push the issue, but rather than turning to government programs he took a new tack, pledging to enlist churches and synagogues—"faith-based programs"—as front-line warriors in his GOP version of the war on poverty. By advocating seed-money grants to churches to help them set up federally-funded programs for the poor, Bush proposed to shift the locus of antipoverty efforts from the public to the voluntary sector. The Bush plan included a big expansion in tax incentives for charitable giving, permitting taxpayers who did not itemize their deductions to list their contributions to charity and get an extra write-off anyway.

The Bush campaign's twin focus on education and poverty came to symbolize its strategic appeal to the center. Whenever polling reflected that Bush had ventured too far to the right in his effort to secure his Republican base, he would return to his education and poverty messages to strengthen his new foothold in the nation's ideological middle ground.

But Bush knew he couldn't win simply by appealing to the center. He also had to satisfy traditional conservatives. The key to his strategy was his ability to balance his centrism on education with a hard-right position on the tried-and-true GOP issues of abortion, taxes, and defense.

Any company or group that seeks to update its image or reach new markets must take care not to let go of its base. The lesson of a success like Bush's is that he was able to retain the loyalty of his traditional partisan supporters even as he acted to expand his party's reach to the center.

As Bush walked the tightrope of presidential politics, he balanced himself like a circus performer with two barbells. On the left, he used education. On the right he used traditional Republican issues. When he swayed too far in one direction and threatened to fall, he would regain his balance by using the other issue to stabilize himself. By this strategy of balance, Bush was able to move his party to the center without losing the backing of its orthodox and passionate right wing.

Conservatives were generally inclined to trust Bush, more than they had trusted his formerly pro-choice and moderate father, because of his record on crime in Texas. As governor, Bush resolutely supported his state's death penalty. After 111 men were executed during his six years in office, he had to consider whether to allow the execution of a woman, Karla Faye Tucker. Bush's response won him lasting traction on the right. By rejecting repeated petitions for clemency, Bush allowed the execution. Demanding that people who commit crimes should suffer the consequences, and noting that he had "sought guidance through prayer," he refused to stay her execution.

In the Michigan GOP candidates' debate, when moderator Tim Russert asked Bush "And what would Jesus think of the death penalty?," the candidate's response was unwavering: "I support the death penalty because I believe it saves lives . . . I believe it protects innocent people to have the death penalty." He added, "I'm a lowly sinner myself; I'm not going to put words in Jesus's mouth."

The race for the 2000 Republican nomination had begun with a strong focus on tax cuts. As each candidate presented his own plan in the televised debates, the contest began to sound like a bidding war, with each campaign attacking the others for proposing tax cuts that were too small. In a move to end the battle and prove his mettle, Bush outbid them all with a $1.6 trillion tax cut. Time after time, he insisted that nobody should have to pay more than a third of his or her income in taxes. Bush told the Republican Convention and the nation that "we will reduce tax rates for everyone." By cutting rates for lower, middle, and upper income brackets alike, Bush made his tax cut sound downright populist: "The surplus is not the government's money." he said. "The surplus is the people's money."

For good measure, Bush also proposed elimination of the estate tax—dubbed the "death tax" by GOP publicists—and the repeal of the so-called marriage penalty, which taxed married couples at a higher rate than it would two unmarried people with identical incomes. Bush also remained a doctrinaire, hard-line, pro-lifer on abortion, though he chose not to engage the issue, hinting that the time was not right for major progress.

Bush completed his right-wing trilogy with a dogmatic determination to strengthen the military after a perceived eight-year period of decline under Clinton. Stressing his support for higher defense spending in his acceptance speech, Bush declared that "America's armed forces need better equipment, better training and better pay." Attacking the lack of military readiness and morale, he pledged "a billion dollar pay raise for the men and women who wear the uniform." He stressed the need to "rebuild our military power," and supported the development of antiballistic missile systems "to guard against attack and blackmail."

Not only did Bush's positions reassure the party faithful of his fidelity but his eager embrace of the Republican iconography sent the same signals as well. Ronald Reagan remained George Bush's lodestar. Paying obeisance to a man who had, in what now seems another era, defeated his own father for the presidency in 1980, Bush the Younger affirmed his true party loyalty.

Iconography has always mattered greatly in American politics. In this century Democratic campaigns have become virtual history classes, trotting out the images of FDR, Wilson, Truman, and, above all, the beloved John Fitzgerald Kennedy to rally support for the latest pretender to their throne. Until recent years, in contrast, the GOP has suffered from an icon gap: Lincoln is too remote, Theodore Roosevelt too liberal, Eisenhower too passive. About Herbert Hoover and Richard Nixon, of course, the less said the better.

But the ascension of Ronald Reagan to legendary status allowed the Republicans to indulge in the politics of ancestor worship with an enthusiasm that matches the heights of JFK idolatry. George W. Bush took time early in his campaign to sing the praises of his father's one-time boss, calling Reagan "a man of deep convictions and a great sense of humor, a kind and decent man of principle."

But Bush didn't stop there. Along with invoking the icons of the past, he worked to bolster his GOP credentials throughout the campaign, arranging a weekly stream of endorsements from Republican governors, senators, and congressmen. He raised funds from all of the Republican special interests, each donation bringing not only money, but a stamp of approval from the right wing.

Thus reassured, the right was able to look on benignly as Bush courted the moderates.

As his campaign unfolded, Bush would use education to win women over to his side, while his messages on taxes, abortion, and defense issues attracted men. He focused on schools to win younger voters, and his trio of Republican issues to get older ones.

From the moment the campaign season began Bush tacked from right to left to right again, fighting off challengers on both flanks.

At first, he introduced himself to America as a moderate. In announcing his candidacy and lining up early support, Bush worked hard to use his education and antipoverty ideas to convince skeptical voters that he was qualitatively different from orthodox candidates like Bob Dole and Newt Gingrich.

But as he began to face Republican challengers in the pre-primary debates, he shifted his emphasis to his ambitious tax cut program to build support on the right. When Arizona senator John McCain upended Bush in the New Hampshire primary, the Texan suddenly found himself fighting for his life.

What went wrong? For one thing, he may have been pushing his hard-right agenda too much. In New Hampshire, both Republicans and Independents are allowed to vote in the GOP primary. When the smoke cleared it became apparent that, while Bush had won over-whelmingly among Republicans, he lost by equally lopsided margins among Independents.

It was a revealing lesson, but one he couldn't dwell on for long. Rather than move to the center to compete for Independents, Bush knew he had to stay to the right a little longer, as he rounded up the true believers and pushed onward to the next primary state, South Carolina.

In this conservative stronghold, which the iconoclastic moderate McCain was foolish to contest, Bush held his thumb firmly on the most emotional right-wing hot buttons, attacking McCain for being soft on abortion. McCain was staunchly pro-life, but campaign finance reform was his core issue—and among his proposals was one to limit soft money contributions. Inventively, the Bush people seized on the issue and claimed that McCain's idea would weaken the power of pro-life organizations, giving the left a huge advantage. Bush also firmly rejected any effort to modify the GOP's tough position on abortion in its platform. The strategy paid off. Bush enjoyed a 53-42 victory over McCain in South Carolina. But along the way he had taken a risk. Had he alienated voters who had originally been attracted to him as a moderate?

Confronting the threat that he might lose his balance and fall off the presidential tightrope, Bush used the barbell on his left to regain balance. Once McCain was defeated, Bush lost little time returning to his education and poverty agenda as he prepared to take the national stage at the GOP convention. By showcasing Powell and Rice, emphasizing education, and devoting his prime-time acceptance speech largely to compassion and opportunity, Bush worked to reassure those who felt he had strayed from the center.

As he entered the battle with Al Gore during the general election, Bush sometimes seemed to be playing an elaborate version of "pin-the-tail-on-the-donkey," designed to attach himself to that particular part of the Democratic symbol's anatomy by following Gore around and echoing his messages.

His opponent staked out his priorities at the Democratic National Convention with a remarkably detailed exposition of his agenda. Sounding more like a state of the union address than a speech to partisan supporters, Gore's acceptance address spelled out exactly how he proposed to further the successful Clinton-Gore agenda in the new century. Polls showed that the speech vaulted Gore into a narrow lead over Bush, overcoming the seventeen-point deficit with which he had entered the convention.

But in the days that followed, an odd thing happened. Wherever

Gore went, Bush went too. Gore urged support for a patients' bill of rights to allow those who were denied adequate treatment by HMOs to sue to get proper care. Bush answered with support for the watered-down GOP patients' bill of rights. Gore backed a new prescription drug benefit for the elderly under Medicare, remedying a hole in the safety net that was particularly difficult for seniors who were poor and sick. But Bush came right back to second the motion, embracing a Republican version of the idea. When Gore promised action on education, Bush was right there too, pushing his own education plan.

With his frustration evident, Gore entered the debates with his Republican rival out for blood. Offended and upset by Bush's pretending to embrace the Democratic agenda, Gore tried to distinguish between his plan and Bush's on each of his issues. The result was a morass of detail that bored voters. The more trivial the differences seemed, the more Gore tried to pound them home. In the end he succeeded, not in making Bush seem too right-wing, but in making himself seem pushy, obnoxious, and even boorish.

By election day, Bush—the right-winger of the primaries—had regained his balance in the center.

Undoubtedly, George W. Bush won the presidency because he triangulated, changing the message of his party to attract independents and convert Democratic issues into areas of Republican strength. But before Bush was able to win the election by changing his party, he first had to secure his party's nomination by pledging allegiance to its essential message.

By building his partisan foundation in the primaries, then moving to the center after the convention, Bush was able to retain the support of his base while he prospected for votes in new territory. His tax cuts, pro-life position, support for defense spending, adoration of Reagan, strong support among party leaders, and focus on using programs of the right to solve problems of the left, all combined to send a message to his base: "I have to leave on a business trip," he seemed to be saying to his hard-core Republican family. "I must, after all, succeed in the world. But I'll be back home before you know it, and I'll bring lots of presents with me."

This conscious balancing act is vital for anyone seeking to broaden the reach of his party, business, or civic group to include a different philosophical base. One must alternate a commitment to the traditional with a determination to embrace the new, applying first one and then the other to stay balanced on the tightrope.

If Bush had found a formula for keeping his party in line as he led it in a new direction, it could be argued that he learned the technique from his immediate predecessor. What Bush did so well, Bill Clinton virtually invented.

BILL CLINTON LEADS
HIS PARTY TO THE CENTER

Moving to the center is not as easy as it may sound. Any political candidate who tries it, whether he comes from the right or from the left, must battle the gravitational pull of his party's traditional ideology as he seeks to move away from its orthodox positions and embrace new ones. He must battle this partisan gravity with a determined centrifugal force as he tries to stake out his independence and develop his own third way.

Departing from a party's traditional ideology is a bit like kicking the bottle. Just as an alcoholic faces a lifelong conflict between his inner thirst and his learned need for restraint, so a politician attempting to reposition himself must fight the forces of conformity and party discipline and daily assert his determination to blaze a different trail.

Bill Clinton fought an eight-year battle against the gravitational pull of his own party's liberal wing. As he ran for president, Clinton hugged the center in order to win the votes of Independents. On taking office he moved left, pushed by a liberal Democratic majority in both houses of Congress. After the Republicans took Congress in 1994, he moved back to the center to enable himself to govern in the new climate and secure reelection. In the end the strategy worked, though he would close his days in office forced to the left by his dependence on the Senate Democrats who helped keep him in office after the Monica Lewinsky affair.

To trace these gyrations is to understand how difficult it is to shed the conformist and orthodox straitjacket into which each party seeks to stuff its candidates and officeholders. Every time Clinton's internal gyroscope called for a move to the center, the ideological police of his party were there to try to drag him back to the left. That he could defy them enough to balance the budget, cut crime, and slash welfare rolls is a testament to his courage and skill. That he failed to stay in the center during his second term—and again became his party's slave—is a sad reminder of his personal vulnerability.

The challenges Clinton faced in repositioning himself offer a case study for any organization seeking to move its branding or change its image. The forces of orthodoxy are potent and not easily defied. Like Bush, Bill Clinton found it necessary to alternate periods of independence and fealty, tacking one way and then the other as political needs changed. Such gymnastics are not the exclusive preserve of politics, but are standard operating procedure for anyone seeking to update his company's or organization's posture.

As he conceived his candidacy, Clinton was determined to change and move his party to the middle. Defeat had become habitual for the Democratic Party. After controlling the White House for twenty-eight of the thirty-six years from 1932 to 1968, it suddenly found itself losing election after election. From 1968 to 1992, the party held the presidency for only four out of twenty-four years, and watched Jimmy Carter (1980), Walter Mondale (1984), and Mike Dukakis (1988) go down to devastating defeats. The only presidential contest it managed to win was the post-Watergate election of 1976, and even then Jimmy Carter defeated the unelected incumbent, Gerald Ford, only narrowly.

As Arkansas governor Bill Clinton pondered his race for the presidency in 1992, he knew—as surely as Bush realized eight years later—that his party had to change. He saw that a basic reorientation of his party's positions was essential if he were to have any chance of winning.

In Franklin Delano Roosevelt's time and for decades thereafter, Democrats had won elections through economic populism—pitting the middle and blue-collar classes against the wealthy constituents of the Republican Party. The turning point in the GOP's fortunes came during the Nixon years, when it learned at last how to use social populism to fight economic populism.

While Democrats railed against the economic elite, Republicans attacked the social elite. For the Democrats, the beast lived on Wall Street. For the Republicans, he dwelled in Washington, Harvard, and Hollywood. The Democrats fought economic exploitation by business, while the Republicans attacked overtaxation by the government. Stoked by racial fears, concerns about crime, a reaction to immigration, and an evangelical commitment to values, the Silent Majority of Richard Nixon and the Moral Majority of Jerry Falwell beat back the New Deal Coalition of FDR, JFK, and LBJ.

To offset the Democratic promises of housing, jobs, education, health care, help for the elderly, and environmental protection, a new Republican agenda emerged. Reducing taxes, prohibiting abortion, protecting gun owners, toughening criminal sentences, using the death penalty, keeping immigrants out, cutting welfare rolls, fighting drugs, and protecting family values against media-induced liberalism animated their surge to power.

The demise of the Democrats seemed complete when Republican Ronald Reagan developed an economic agenda to add to the social issues already in the GOP arsenal. By focusing on tax reduction, cuts in government regulation, and supply-side stimulus, Reagan launched America on a decade of prosperity. Ever since Roosevelt rescued America from the Great Depression, prosperity had been a Democratic issue. Now, even that was gone.

The combination of a new Republican economic front and its conservative social populist agenda proved too much for the Democrats to handle. In 1984, Walter Mondale sealed his doom after promising to raise taxes if elected. In 1988, Mike Dukakis lost after defending his opposition to the death penalty and the granting of a weekend furlough to rapist Willie Horton from a Massachusetts prison. (Horton had used his two-day liberty to rape again.)

By 1992, the Democratic Party's confidence was shattered. Even its stalwarts conceded that repositioning was vital.

Bill Clinton had watched this parade of Democratic debacles throughout the 1980s, as he nursed his own presidential ambitions. Working with moderate colleagues in the party, he helped form the Democratic Leadership Council (DLC), dedicated to moving the party back to the center.

Insisting that the traditional debate between the Democrats and

Republicans was obsolete, Clinton embraced the idea of a third way that would reject the orthodoxies of each camp.

Clinton articulated this paradigm:

"Now our new choice plainly rejects the old categories and false alternatives they impose. Is [it] liberal or conservative? The truth is, it is both, and it is different. It rejects the Republicans' attacks and the Democrats' previous unwillingness to consider new alternatives."

Frankly criticizing the Democratic liberal orthodoxy of the 1980s, he called for a new focus on the problems of the middle class, eager to bring them back home to the Democratic Party after their years in the Nixon/Reagan social-populist wilderness. "Too many of the people that used to vote for us," he said, "the very burdened middle class we are talking about, have not trusted us in national elections to defend our national interests abroad, to put their values into our social policy at home, or to take their tax money and spend it with discipline. We have got to turn these perceptions around, or we cannot continue as a national party." Clinton believed the nation was enervated and frustrated by the endless ideological debates in national political life. He rejected the simple assumptions of each side—the Democratic view that government was the answer, and the Republican conviction that it was the enemy. He felt that government had its place, as part of a mosaic of public and private actors that could work in concert to address the nation's problems.

Clinton demanded that the Democrats adopt a course that combined "opportunity, responsibility, and a belief in community."

By embracing "opportunity for all," Clinton was pushing a long-term Democratic agenda that promised upward mobility to its low-income supporters. In advocating "responsibility from all," he sought to reach across the partisan divide and engage with traditionally Republican issues like fighting crime and reforming welfare. His proposed "community of all" was meant to recast government as a catalyst, stimulating all the sectors of society—voluntary, public, and private—to empower people to change their own lives.

But as with Bush—or any company or group seeking to reposition itself, certain signature issues must signal such a transition. Clinton chose three: crime, welfare, and taxes.

Until Clinton, Democrats had treated any discussion of crime as though the issue were nothing more than a code for discussing racism. Rejecting the idea that crime was a federal issue (just as Republicans in the '90s insisted that education was only a local question), they opposed capital punishment and backed the liberal rulings of the Supreme Court.

From the start, Clinton behaved differently. Oddly enough, his first challenge, as with Bush, arose over the death penalty—the case of Ricky Ray Rector during the 1992 campaign. Rector, who was brain-damaged, had murdered a police officer. To many he was a poster boy for the death penalty's inhumanity. Traditional liberals pressured Clinton to grant clemency, but Clinton refused.

Only four years before, Democratic presidential nominee Mike Dukakis had dealt himself a serious blow over the issue, when he failed to show any emotion when asked whether he would back the execution of someone who had raped and murdered his wife. Dukakis's recital of statistics may have suggested that executions did not deter murder, but his tepid demeanor suggested a serious failure to grasp voters' strong feelings on the subject.

Mindful of Dukakis's example, Clinton refused to grant clemency to Rector. In addition, he interrupted his campaign to return to Arkansas and deny clemency to Steven Douglas Hill, a twenty-five-year-old convicted murderer. The *Washington Post* noted at the time that the cases were "often cited to bolster Clinton's claim that his position on the death penalty and other issues makes him a 'different kind of Democrat.'"

Perhaps more memorably, Clinton also departed from Democratic orthodoxy in his campaign by running a TV ad promising to "end welfare as we know it." Demanding that welfare recipients work for their benefits, Clinton sounded more Republican than Democratic at the 1992 convention when he declared that "you have a responsibility to go to work. Welfare must be a second chance, not a way of life."

Democrats had come to regard any suggestion of ending welfare entitlements as tantamount to racism. Whenever the issue was raised, liberals would nod knowingly and counter, "We know what you're trying to do. You're attempting to use racial prejudice to get elected. And you're speaking code words to do it."

But Clinton insisted on decoding the book. He demanded that we scrutinize the lives of the welfare poor, and judge honestly whether the system really helped them or only trapped them into a dependent "way of life." He emphasized not the venality of the impoverished, but the glories of self-respecting work, as he pried welfare reform loose from the grip of conservatives and made it a Democratic issue.

Clinton completed his own realignment hat trick by committing himself to cutting the federal budget deficit in half by the end of his first term. Despite his robust history of tax hikes during his tenure as governor of Arkansas, Clinton tried to distance himself from what Reagan liked to call the "tax-and-spend-Democrats," pledging his support for a middle-class tax cut.

Cloaking his freshly assembled ideology in Christian tones, Clinton called his approach a "New Covenant." The era of big government and free lunches, of "coddling tyrants" and criminals, was over. Clinton was a "new Democrat."

Clinton demonstrated his move to the right in his 1992 campaign by attacking the black rap singer Lisa Williamson, a.k.a. Sister Souljah, when she was quoted as saying, "If black people kill black people every day, why not have a week and kill white people?" Clinton snapped back that "if you took the words 'white' and 'black' and reversed them, you might think David Duke [the Louisiana white supremacist] was giving that speech."

With Ross Perot winning 19 percent of the vote as an Independent, Clinton was able to take the White House in 1992 despite capturing a mere 43 percent of the national vote. Polls indicated that about 35 percent of the votes came from traditional Democrats— liberals and minorities who backed his candidacy solidly. But the crucial remaining eight points came from moderates attracted by the possibility that Clinton was, indeed, a new kind of Democrat. It was a classic triangulation victory.

But while occupying the White House, Clinton seemed to succumb to his party's gravitational pull. With the Democrats keeping control of both houses of Congress, the new president listened hard to the appeal of Senate Majority Leader George Mitchell and House Speaker Tom Foley for party unity and discipline.

Their siren song was hypnotic. Pledging their undying loyalty, they appealed to the new president—a young outsider with limited Washington experience—to work in tandem with the Democratic Congressional caucus. Well aware of the way Democratic Congressional majorities had torpedoed the last "outsider" Democratic president, Jimmy Carter, Clinton was determined not to repeat the mistake.

As Clinton looked across the party aisle at the Republican congressional delegation, the prospects for bipartisan cooperation were not appealing. Treating Clinton's victory as a fluke caused by the Bush–Perot split in GOP ranks, the Republicans vowed to wait out his presidency to reassert their claim to the White House. Indeed, almost as soon as Clinton took office, Senate Republican Leader Bob Dole announced his likely candidacy against him in the next election.

The dilemma Clinton faced as he assumed office is not unlike that of any new CEO or president of an established group. The path of the conventional beckons with its comfort, company, and comradeship. It is not your enemies who seduce you into conformity—it is your friends, your would-be allies.

Foley and Mitchell, the Democratic congressional leaders, offered Clinton the comforts of the familiar. "Why do you want to risk alienating us, your best friends? Stay with us. Stay in the fold. We'll protect you. We'll nurture you. We'll meet your every need. Why venture out into the cold? Come inside, make yourself at home," they seemed to say. But in accepting their invitation, Clinton must have realized that it was, indeed, a prison.

Turning one's back on supporters isn't easy. Looking away when they counsel conformity is even harder. Yet, as Clinton's experience shows, it is sometimes irresistible.

Rejected by the Republicans, Clinton seemed to abandon his centrist orientation and accept the liberal embrace. He began his administration by announcing his support for allowing gays to serve in the military. Legislatively, he sent Congress a stimulus package of increased federal spending and a tax increase, as he tried to catalyze the economy, reduce the deficit, and lower interest rates. And he drifted further to the left by appointing First Lady Hillary Rodham Clinton to revamp the nation's health care system by creating a system of nationwide managed care and universal health insurance coverage.

It was a series of missteps that infuriated Republicans and alarmed his new Democrat coalition almost equally.

As Clinton dragged his feet on his two key campaign promises—a middle-class tax cut and welfare reform—voters turned against him en masse, feeling betrayed by the president's turnabout. As he entered his second year in office, Clinton's approval ratings had dropped to the low forties.

With his popularity decreasing, Clinton found himself unable to attract the support of moderate Republicans in Congress, and was increasingly forced to depend on Democratic party loyalty in passing his program. With each passing week, the party's gravitational pull on him moved Clinton further and further left.

Even when he tried moving back to the center, the new president's luck went against him. In 1993 Clinton proposed a strong anticrime bill; while it embraced liberalism by urging strong gun controls, the plan also expanded the grounds for federal capital punishment, proposed hiring 100,000 extra police, and sought massive prison construction. The black and Hispanic caucuses were aghast at the proposals. When Democratic House leaders pressured the groups to vote for Clinton's bill, they were forced in return to promise new spending on inner-city programs like midnight basketball courts and swimming pools. What started out as a conservative anticrime bill became, in public perception, a big spending giveaway to minority neighborhoods.

When his anticrime bill only barely managed to squeak through Congress in August 1994—months before the Congressional elections—Clinton seemed to be in trouble.

Hillary Clinton made matters worse with her health care program, which conservatives argued would restrict patients' ability to see their own doctor. When her proposals died in committee in early September 1994, Clinton faced a major reversal in the midterm elections. Though he had anticipated a setback, Clinton was shocked to find himself buried beneath a landslide, losing both houses of Congress to conservative Republicans under their fiery new speaker, Newt Gingrich.

Why did Clinton move so far to the left? He'd never wanted to. But the pressure of his party's congressional leadership, and his cold rejection by the conservative Republicans, left him feeling that he had no choice.

Clinton's first steps to the left at the start of his term—his gays-in-the-military position and his support for a stimulus package and tax increase—were like an alcoholic's "just one drink." Gravity took over. His early moves to the left cut off his nascent support from the center; that, in turn, forced him to turn ever further to the left in search of congressional approval. This increased liberal tilt led to more rejection in the center. As Clinton told me, "I became so liberal I didn't even recognize myself."

Perhaps there is something inevitable about this postelection tendency to return to party orthodoxy. During his first six months in office, George W. Bush made the hard-line conservative agenda his first order of business, slashing taxes, rejecting the treaty against global warming, seeking to roll back Clinton administration protections against arsenic in drinking water, opening the Arctic Wilderness Preserve to oil drilling, and moving to promote a missile defense shield. Bush's ratings dropped five to ten points as he entered his second six months in office; only after he took the opportunity to highlight a few more moderate positions—allowing federal funding on stem cell research to continue, strengthening controls on water pollution, and shelving his voucher program in favor of a big hike in education funding—did his numbers start to rise.

But Clinton's troubles in 1994 were far worse than anything Bush has yet faced. With Republicans in charge of Congress, Clinton had only one option before him: playing off the GOP agenda as surely as he had in 1992, he began once again to move to the center.

Confronting the disastrous midterm election results, Clinton realized that he had no choice but to abandon old-school liberalism. He had learned in 1993 and 1994 that the Republicans could stand up to his opposition; now, in 1995 and 1996, he would discover gladly that they were less skillful in handling his cooperation. The Republican Party, it turned out, could not take Clinton's yes for an answer and survive politically.

Clinton's opening salvo came in his State of the Union address, when he declared that "the era of big government is over." Clinton announced his change of direction with a pledge to cut spending, hold down taxes, and eliminate the deficit.

Before long it was clear that a grudge match between Bill Clinton and Newt Gingrich was taking shape. Upon taking power, Gingrich moved aggressively to slash federal spending and pass the right-wing agenda. Committing himself to a "Contract with America," Gingrich sought to roll back environmental regulations, dismantle the federal Department of Education, and scale back Medicare and Medicaid. Laying out a specific program of budget cuts, the Republicans dared Clinton to propose an alternative of his own.

In perhaps the most dramatic instance of Clinton's triangulation, he did just that. In a national prime-time TV address in June of 1995, he announced his own plan to eliminate the deficit within ten years.

Clinton said his budget had "five fundamental priorities": preserving education funding, controlling health care costs, cutting taxes for the middle class, fighting crime, and reducing welfare. Clinton claimed that "for twelve years, our government . . . ducked the deficit," and challenged the country to do better: "In my first two years as President, we turned this around and cut the deficit by one third. Now let's eliminate it. It's time to clean up the mess," he continued. "This debate must go beyond partisanship. . . . If we'll just do what's best for our children, our future and our nation and forget about who gets the political advantage, we won't go wrong."

His proposal represented a historical shift for the Democratic Party. Although Democrats had always paid lip service to ending the federal deficit, and cried loudly when Reagan had more than tripled it, they had cautiously avoided ever proposing an actual balanced budget themselves—or even a glide path to one.

Clinton's repositioning on the issue gave him new legs to stand on as he fought the harsh cuts Gingrich sought in federal spending. Republicans had expected that they could sell their spending cuts as essential to eliminating the deficit. Instead, Clinton had shown that such dire reductions were not necessary. Republicans suddenly found themselves defending their drastic plan against the more gradual path of deficit reduction proposed by the president.

When Republicans sought to force their budget through by shutting down the federal government twice in late 1995, Americans sided with the president and rejected the GOP proposals. By 1997, when he negotiated a budget deal with chastened Republicans that closely paralleled his initial proposals eighteen months before, Clinton would finally be vindicated.

Facing a tough battle for reelection in 1996, President Clinton had further bolstered his moderate credentials by signing historic welfare reform legislation three months before election day. The new law required all able-bodied welfare recipients to work in order to collect checks, and set a five-year time limit for welfare dependency.

Twice before, a GOP Congress had passed similar legislation only to see Clinton veto the bill. But previous GOP efforts at reform had slashed nutrition programs, cut protective services to prevent child abuse, and failed to increase day care to enable welfare mothers to work. The Republicans had also attempted to cut the Medicaid program, which provided health care to the welfare poor.

Daring Clinton to veto the bill, the Republican Congress passed its welfare reform package and placed it on the president's desk. While Clinton approved of the bill, he agonized over whether or not to sign it because it included huge cuts in benefits to legal immigrants. "If a man comes here with his family legally," he argued to me, "pays his taxes and works every day and gets injured, why shouldn't he get disability payments just like the rest of us?"

The debate split the Democratic Party down the middle. The welfare reform law had passed the House of Representatives, with ninety-nine Democrats voting yes and ninety-nine no. Would Clinton sign it? The direction of his party hung in the balance.

Gambling that he could reduce the cuts in immigrant benefits in his second term, Clinton affixed his name, removing a crucial element in the GOP appeal to the frustrated middle class. And his luck held: In 1997, largely under pressure from Republican governors whose states would be forced to bear the entire tab for immigrant benefits themselves, Congress repealed the cuts—saving Clinton the trouble.

As Clinton's second term unfolded, the effectiveness of his plan was proved beyond a doubt. Each month saw a drop in welfare cases throughout the nation. By the end of his administration, the welfare population had declined by 40 percent nationally—a virtual revolution in the lives of America's poor. The long-term, generation-after-generation welfare cycle of dependency had been broken—by a Democratic president!

By eliminating the deficit, reducing crime, and slashing welfare, Clinton had accomplished everything middle America could have wanted from a president of either party. His efforts had moved the Democratic Party to the center, and erased its image as soft on crime, profligate on welfare, and irresponsible on the deficit.

Yet the most visible triangulation in Clinton's agenda was his shift from an emphasis on economics to values as the core of his program during his campaign for reelection.

Clinton was determined to recapture the moderates the Democrats had lost in the '70s and '80s to the social populism of Nixon and Reagan. Focusing on values issues, Clinton's platform highlighted a broad program aimed at dealing with family problems. He called for stricter child support enforcement, background checks for child care workers, gun controls on school grounds, a ratings system to warn parents about violent or sexual TV programming, tougher enforcement of drunk driving laws, expanded family and medical leave, and a host of other values-based programs.

Clinton's emphasis on values issues stemmed directly from a gaping hole in his political appeal. While polls showed him decisively defeating his Republican rival, Bob Dole, among single voters and married couples without children, he lost to the Republicans among voters who were parents with children living at home.

Indeed, five basic values questions emerged as a highly accurate predictor of whether someone would vote for Clinton or for Dole. Those who gave predominantly conservative answers voted for Dole, while those whose answers showed that they were less tied to traditional values voted for Clinton.

The questions were:

1. Do you believe sex before marriage is morally wrong?
2. Is religion very important in your life?
3. Do you ever personally look at pornography?
4. Do you think less of someone because they cheat on their spouse?
5. Is homosexuality immoral?

Had the five questions been about labor unions, income redistribution, poverty, taxation, and crime, their predictive value would not have been so surprising. But for social-values issues to hew so closely to partisan divisions deeply disturbed the president. That these questions would prove to be such accurate predictors of voting intention underscored, for Clinton, the fundamental importance of social-values issues.

Clinton, whose internal 1992 campaign motto was "it's the economy, stupid," now called on his administration to focus on an agenda more social than economic. The new strategy was on display prominently in his 1996 State of the Union message, as the president focused on challenges to America ranging from teen smoking to drug use to illegal immigration to limited college opportunities, along with a legion of other social values issues.

In Clinton's second inaugural address he continued to look toward the center, emphasizing traditionally conservative issues like crime and welfare reform. "As times change," he said, "so government must change . . . it can stand up for our values." He called for "hiring people off welfare," and stressed that "each and every one of us, in our own way, must assume personal responsibility," so that "our land of new promise will be a nation that meets its obligations—a nation that balances its budget, but never loses the balance of its values."

Clinton's move to values issues mirrors the growing tendency of major corporations and groups to shed direct consumer appeals in their advertising in favor of more subtle tactics. Southwest Airlines doesn't fly planes anymore—it offers "freedom." A telephone company doesn't just offer phone service—it helps you "reach out and touch someone." The transition from products and services to values is endemic to much of twenty-first-century American life.

The lesson of Clinton's repositioning toward values lies in the importance of building toward the transition with a lot of relatively small steps, rather than trying to bridge the gap with a few huge ones. The bridge toward a values-based agenda must consist of many individual bricks, each contributing to the overall theme or impression. At first such an incrementalist approach may be open to criticism as "small-bore." But in an era when big government is dead, the same may have happened to big solutions. It is by a series of small steps that one must often make progress.

While Clinton was moving to the center, how did he retain the support of his own party? As he embraced issues like balancing the budget, cutting crime, and reforming welfare, after all, he ran a serious risk of alienating his own party and losing their support.

It wasn't easy. When Clinton announced his plan to balance the federal budget in June 1995, he was met with withering criticism from the Democratic ranks. Congressional Democrats, anxious to attack the Republican Party's proposed budget cuts, felt the president's speech undercut their focus on the Medicare reductions the GOP was urging. As soon as Clinton ended his ten-minute speech, criticism began pouring in from Democrats on Capitol Hill. According to Clinton aide George Stephanopoulos, neither Richard Gephardt, the House Democratic leader, nor Tom Daschle, the Senate Democratic leader, liked the speech. The *New York Times* reported that senior Democrats were "unsparing in their criticism," and Rep. Nancy Pelosi, a Democrat from California, said Mr. Clinton was "playing right into the hands of the Republicans." Gephardt said that Clinton had "turned Medicare into a political football," adding that "the real losers will be the elderly and the families that support them." When the Clinton deficit-reduction plan came up for a vote in the Senate it was rejected 99-0, as the Democrats joined unanimously with the Republicans in repudiating the president's position.

With such aggressive criticism, how was Clinton able to unite the Democratic Party behind him as he transformed it from a liberal to a centrist political force? How did he straddle the gap between the parties without falling on his face in the middle?

As with Bush, the key lay in what he did *not* change in the Democratic agenda. Just as Bush followed traditional GOP doctrine on taxes, abortion, gun control, and a host of other issues, so Clinton never abandoned Democratic orthodoxy where it counted.

On abortion, Clinton stayed solidly pro-choice, even vetoing legislation to ban third-trimester abortions—the controversial procedure also known as partial birth abortion, which seemed to many to resemble infanticide.

Indeed, Clinton pushed to the forefront a strong defense of Medicare, education funding, and environmental protection—three key elements of the Democratic agenda. He was vociferous in attacking Gingrich's proposals to slash funds for environmental protection and to scale back dramatically the rule-making authority of the Environmental Protection Administration. Most important was Clinton's opposition to Republican proposals to cut Medicare, the program that provides medical services to the elderly. The Republicans sought to cut the growth of the program by almost $300 billion, increasing the deductibles and co-payments that the elderly would have to pay.

Rather than submit to a budget that included these cuts in

Medicare, education, and the environment, Clinton twice vetoed GOP budget bills—and in retaliation the Republicans took their bill and went home, shutting down the government for lack of funds. With all but essential federal agencies closing their doors, Clinton hardened his position and let voter resentment against the Republicans and their brinksmanship mount.

Running television commercials attacking the Republican budget cuts, Clinton argued that he could balance the budget without reductions in these vital programs. Voters embraced the Clinton message, and each day that the government agencies were shuttered, the Republican negatives in national polls grew.

When the GOP finally caved in and reopened the government, while meekly rescinding its budget cuts, Clinton scored a resounding victory. His win solidified the Democratic Party behind him, even as he ventured into untested waters by signing the welfare reform bill.

Clinton also kept the Democrats on the reservation by appealing directly to the African American constituency, which underscored the Democratic base. Just as Bush toed the line of the Christian Coalition on abortion and other social issues, Clinton refused to compromise in his support of civil rights programs the black community wanted.

The key test of Clinton's resolve on the subject came when California passed a 1994 initiative to repeal affirmative action. Attacking the program as discrimination in reverse, conservatives succeeded in eliminating preferences for minorities in hiring, government contracts, and university admissions. Though he faced massive pressure to abandon affirmative action, Clinton refused; he backed the program on principle, calling to "mend it, but don't end it." Supporting token reforms like a ban on racial quotas and a provision that no white lose his job due to affirmative action, Clinton resisted calls to backpedal on this longtime Democratic priority. Minority and women's groups hailed the president's decision; they would stay with him throughout the controversies that came to envelop his presidency.

Even when Clinton did break away from Democratic dogma, he was careful to leaven his new positions with a dash of the more traditional and liberal elements of his party platform. While his anticrime initiative made much of support for the death penalty and expansion of prison capacity, it also focused on gun controls dear to the hearts of Democratic liberals. Clinton's welfare reform proposals always included expanded day care for welfare mothers, and job training to help them cope with the demands of employment. His bill also included massive tax subsidies to encourage employers to offer jobs to those formerly on welfare. Anytime Clinton focused on retiring the deficit and balancing the budget, he was careful to note the importance of preserving priority Democratic programs like Medicare, the environment, and education. He strayed far enough to bend—but not to break—the party's traditional liberal core. It is a crucial element in successful triangulation: Be careful to leave some things unchanged, even as you embark on a course of change.

Clinton also used his fund-raising prowess to help bridge the gap between him and his party's elders. Plunging into fund-raising with a vigor and effectiveness unique in presidential annals, Clinton raised over $300 million for his 1996 reelection race and helped gather hundreds of millions more for Democratic House and Senate candidates. Appearing on the campaign trail with even marginal Democrats running for House and Senate seats, Clinton did all he could to atone for his infidelity to party dogma on Capitol Hill.

The lesson from Clinton and Bush is clear: You can move to the center by trying to solve the problems normally associated with the other party, but don't abandon all the traditional positions of your own party on the issues with which it is normally associated.

Clinton's final years in office, of course, were dampened by developments that no amount of political strategy could entirely overcome. Having broken out of his party's orbit from 1995 to 1997, Clinton fell under its sway again as his presidency became crippled by scandal. Once his affair with Monica Lewinsky was unearthed in January 1998, Clinton became the hostage of his party's congressional minority. With Republicans going for his throat, demanding his removal, and then actually passing a bill of impeachment in the House of Representatives, Clinton clung to office by the barest of threads—the one-third plus one that his party could muster in the Senate to block his ouster.

To ensure that his party's senators would stand by him, Clinton was forced to abandon the notion of repositioning. A promising Medicare reform program sponsored by Democratic senator John

Breaux of Louisiana fell at the hands of the party's most hard-line liberals. And by the time of the 2000 election, congressional Democrats were eager to find issues to use in wresting control of Congress from the Republicans. Suddenly any new achievement Clinton could chalk up with Republican support was one less issue for the Democrats to use in running against the GOP in the next election. From granting patients the right to sue HMOs, to giving the elderly free prescription drugs, to reforming Social Security, Clinton now sided obligingly with his party, rejecting potential compromises that might have taken the steam out of important Democratic state and national campaigns.

Often at odds with his party, Clinton ended his presidency giving in to its gravitational pull and ceding his long battle to temper its traditional liberalism. But at his three-year centrist peak, Clinton had accomplished enough to leave behind an important legacy. He had taken a nation beset with high crime, mounting welfare rolls, and a crippling budget deficit and had turned it around by sensing an opportunity to move to the center and triangulating his way to success.

But the strategy of triangulation is not an exclusively American franchise. Indeed, the example I cited to President Clinton in proposing the move to the center came not from our shores, but from France. The story of how the Socialist French president François Mitterrand triangulated his way to reelection against Gaullist leader Jacques Chirac in the 1980s—a move to the middle that saved his presidency—is as revealing as the similar move that saved Clinton's.

FRANÇOIS MITTERRAND LETS JACQUES CHIRAC PASS HIS PROGRAM . . . AND THEN BEATS HIM AT THE POLLS

When President Bill Clinton asked me to help him recover from the Republican congressional victories of 1994, he wondered which historical model he should pursue. Dressed in a T-shirt, jeans, and sneakers as he relaxed in his favorite wing chair next to the couch in his burgundy-colored office in the White House's East Wing residential area, he explained his options. Confronting a Congress controlled by Republicans for the first time since 1953, how was he to govern? "Some people want me to do what Truman did, fight the Congress, confront it," he told me in his urgent southern drawl. "Others say I should act like Eisenhower and try to work with them, give in to them."

He looked at me expectantly as I walked over to his bookcase on the opposite wall. Having thoroughly browsed its shelves during the forty-five minutes I'd spent waiting for Clinton, I took down a copy of a biography of French president François Mitterrand that I'd read a few months before. "Do what Mitterrand did to Chirac," I urged, opening the book and finding the passages I remembered.

Clinton cocked his head at my unexpected answer. I explained how Mitterrand, a socialist, had nationalized vast segments of the French economy when he took power in 1981, only to see France fail economically as conservative administrations in the United States and Britain attracted all serious investment away from Mitterrand's new socialist state. Defeated in the parliamentary elections of 1985, Mitterrand faced a situation no French president had confronted since the birth of Charles de Gaulle's Fifth Republic in 1958—a Socialist president and a Rightist Chamber of Deputies.

"He could have appointed a moderate conservative like Valéry Giscard d'Estaing," I reminded Clinton. But instead he chose to go hard-right and name Jacques Chirac, the leader of the Gaullist RPR Party. With their penchant for the risqué, the French dubbed the Mitterrand-Chirac administration "cohabitation."

Predictably, Chirac privatized all the industries Mitterrand had taken over, reversing the policies that had brought the Socialist to power four years before. Far from protesting, Mitterrand approved all Chirac's legislation, letting the conservative have his head.

"By the time Chirac ran against Mitterrand two years later," I told Clinton, "he had no issues. Everything he wanted to achieve, he already had. There was no reason to vote for him, and he lost. Mitterrand was reelected.

"Fast-forward the Republican agenda," I urged the president. "Cut the deficit, cut crime, reform welfare, solve their problems. Relieve the frustrations that animated their [1994] electoral victory. Then they'll have nothing left to run on in 1996."

In a memo to Clinton in March 1995, I dubbed this strategy "triangulation."

But the question recurs: What was it that Mitterrand was able to do to keep his party in line behind him? Reversing everything that both he and it stood for—it is called, after all, the Socialist Party—he managed still to command its loyalty. How did he pull it off?

The Mitterrand example provides lessons that are, in a sense, even more vivid than those of the Bush or Clinton eras. In triangulating, Mitterrand not only changed positions, he changed roles. He evolved from being leader of a party to president of a nation. He moved from specifics to generalities, and gained authority through the act of ceding power.

In 1981, Mitterrand was a dedicated socialist. Determined to transform the French economy from private to public ownership, he pursued a program of nationalization vigorously, warning that multinational conglomerates would dominate France if the government did not move now to take over key industries. "The nationalizations will give us the tools for the next century," he declared at his first press conference after winning the election—the longest press conference in French history. "If that were not done, instead of being nationalized, these enterprises would speedily be internationalized. I refuse an international division of labor and production decided on far away and obeying interests other than our own. Nationalizations are an arm of the defense of French production." The nationalized economy, he argued, would permit the nation to "pursue just social, economic, and fiscal reforms."

With a majority of the normally fractious French public favoring nationalization of key industries, Mitterrand decided to move fast in transforming France into a state-owned economy. Determined to vest total control in the government, he rejected the advice of moderates who advised that he only take over 51 percent of key companies, insisting instead on 100 percent state ownership. The companies fell under government control at a dizzying pace.

An amazing spectrum of companies were instantly moved from the private to the public sector. Defense industries, including the manufacturer of the Mirage fighter-bomber and other arms producers, as well as electronic producers, pharmaceutical companies, banks, and other corporations, were suddenly placed under government control.

To terrified capitalists, Paris suddenly looked like Petrograd in 1917, as socialism swept all before it.

The immediate results of nationalization were disastrous. French investors reacted quickly and with horror. Private investment dropped 12 percent in 1981 alone. Pulling their money out of France as fast as they could, they fled to the more capital-friendly atmospheres of Ronald Reagan's United States and Margaret Thatcher's United Kingdom.

Meanwhile, nationalization proved to be both expensive and inflationary:

- To take over the companies, the French government had to spend about fifty billion francs ($10 billion)—half for industries and half for banks and other financial companies.
- The new public enterprises received massive government subsidies, costing the nation another 140 billion francs ($28 billion).
- Even with the subsidies the companies operated at a deficit after the takeovers. One expert estimated that in the first three years of nationalization, from 1981 to 1984, the public sector accumulated more than 130 billion francs ($26 billion) in losses.

Nationalization cost 320 billion francs ($64 billion), causing "a serious drain on the Treasury," as Vivien Schmidt's revealing history *From State to Market?* observed. "Within a year, inflation was spiraling upward, and the [French] economy downward. Imports far outdistanced exports, as the high inflation rate decreased the competitiveness of French products . . . the trade deficit with Germany alone grew by 35 percent in 1981, and by 81 percent in the first quarter of 1982." The trade deficit as a whole rose from "fifty-six billion to ninety-three billion francs."

"Profit margins were eroding in the face of rising wage costs combined with the reduction of the work week" as the government predictably moved to mandate shorter hours and higher wages. International banks raised interest rates on loans to French corporations, uncertain of the country's economic future. Stunned, the Socialists did "relatively little to try to counteract the inflationary spiral and the deterioration in the macroeconomic profile of the country that had started almost as soon as they were elected."

As former French cabinet member Henri Weber observed, after their first year in power the Socialists were faced with a political catch-22: "If they accepted reality they had to renounce their socialist doctrine . . . if they rejected reality and stayed with their ideology, they would lose politically." By the mid-1990s the French economy was in freefall.

Socialism had failed.

As the economy dropped, the French left wing found itself facing a disaster in the legislative by-elections of 1986. The rightists pilloried

Mitterrand and his economic policy. One typical French billboard featured a naked woman with a provocative legend UNDER SOCIALISM, I HAVE NOTHING LEFT.

French voters got their revenge that year, ousting the Socialists from control of the Chamber of Deputies and putting the right back in charge. From their ranks, Mitterrand had to choose a new premier.

France's right was divided into two camps—the moderates, under former president Valéry Giscard d'Estaing, and the extreme conservatives headed by future president Jacques Chirac, then mayor of Paris. Either would have commanded the majority of the Chamber of Deputies that a cabinet and its premier needed under France's unique presidential-parliamentary system. Most expected Mitterrand to make the best of a bad situation and choose a member of Giscard's party.

Before 1958 France had had a parliamentary system of government, with a figurehead president and a powerful premier chosen by a majority in the Chamber of Deputies. After years of instability and gridlock, though, Charles de Gaulle had changed all that when he took power in 1958, establishing the Fifth Republic with power evenly, if ambiguously, divided between a directly elected president and a premier chosen by the president from the majority party in the Chamber of Deputies.

The ambiguity of power sharing had never been tested, since the Gaullists and their allies had controlled both the president and a majority in the chamber ever since the Fifth Republic began. The delineation between president and premier remained largely theoretical, the boundary unmarked and untested. But now Mitterrand, a Socialist president faced a conservative majority in the Chamber of Deputies. The prospect of a long and difficult struggle for power loomed ahead.

But, in one of the boldest and most brilliant political moves in modern French history, on March 20, 1986, Mitterrand crossed up the pundits, his advisors, and the bettors by choosing Jacques Chirac to be France's new premier. Passing over compromise choices, Mitterrand handed power to the very man who represented the antithesis of everything he valued.

Why did Mitterrand choose his old Gaullist adversary? It was obvious that Chirac would be Mitterrand's opponent in the coming pres-

idential elections. Why did he choose to give his rival the stature, platform, and power that came with the post of premier?

Some thought Mitterrand was giving Chirac enough rope to hang
himself. Historian Wayne Northcutt noted that Mitterrand knew the
premiership would be "fraught with difficult challenges, such as the
problems of unemployment and terrorism, and that Chirac's popularity with the voters would in time fall as he confronted the realities
of power. The president also knew that Chirac's feisty, aggressive, and
often impulsive manner would cause problems for the prime minister." As Chirac struggled "with the day-to-day problems of [running]
France, Mitterrand realized that he could remain above much of the
political fray and play the role of an arbiter among the right-wing
majority, the opposition and the national interest."

What is clear, in any event, is that he felt confident he could win
the cohabitation battle and defeat de Gaulle's "ambitious, aggressive, and opportunistic political heir," in Northcutt's words, once
and for all.

How often is it possible to win a political battle by giving the other
side what it wants? The humility in realizing that one has erred and
backtracking without stepping aside is hard to come by among
national politicians of any stripe. As with Lyndon Johnson in Vietnam, it can often be more tempting to wade deeper into the swamp
in the hope of coming out with your boots clean, if wet, at the other
end. But Mitterrand was the rare political figure who knew when to
cut his losses and stage a tactical retreat.

Chirac, of course, saw his appointment in a different light—as a
dry run for the presidential elections two years later. Armed with
power, he eagerly proceeded to undo almost all that Mitterrand had
done earlier in the decade. "At the first postelection Council of
Ministers' meeting on March 26, [1986]," Northcutt writes, "Chirac
presented the Right's program, including its plan to denationalize
sixty-five French industries. . . . Thus, the first meeting between Mitterrand and the Chirac cabinet set the stage for the drama of cohabitation."

Chirac's inspiration came from across the Channel and across the
ocean, where Thatcher and Reagan had successfully pitched conservatism and capitalism as the waves of the future. "The transatlantic
example seemed to light the way," Julius Friend writes in *The Long*

Presidency. "Publicists described Ronald Reagan's America in glowing terms—Guy Sorman's *La Révolution conservatrice américaine,* published in 1983, rapidly became a bestseller. Another bestseller, François de Closets's *Toujours Plus* [Always More], criticized trade union privilege, power, and selfishness and raised the question of whether unions were necessary or even legitimate. Other books about the virtues of deregulation, supply-side economics, and less intrusive government rapidly followed." The booksellers of the Left Bank found their stalls full of literature of the Right.

Echoing Rightist demands for less state and more liberty, Chirac proceeded to reverse Mitterrand's nationalization program. But the key question remained: Would Mitterrand stand in the way, or would he step aside and let his new premier and future rival for the French presidency have his way?

The legislation promoting privatization, of course, was popular in the National Assembly. With a top-heavy Rightist majority in the aftermath of the 1986 elections, anything Chirac proposed was likely to receive quick approval. But the Constitutional Council, which the president controlled under the Fifth Republic Constitution, was a different story. Had he chosen to do so, Mitterrand could have had the council declare the privatizations unconstitutional, killing them outright, or at the very least call on them to attach conditions to slow the process down.

Amazingly, the council did neither. Mitterrand let Chirac have a free hand. On June 26, 1986, a bare two months after Chirac took office, the Constitutional Council approved the privatization law. Parliament rushed through the necessary legislation in less than a month, and France was on the way to a massive transfer of economic power from the state back to private hands. Mitterrand's Socialist base could hardly criticize him for going along with Chirac's privatization program. Weakened by its worst political defeat in decades, the Left's true believers were in no position to criticize their president for bowing to popular will and allowing the privatizations. After all, Mitterrand had amply demonstrated his fealty to the Left by going the last mile with his nationalization program, until it almost killed him.

The Socialist Party, like the Democrats in 1992 and the Republicans in 2000, had been chastened by defeat, and their resulting humility gave Mitterrand the same freedom Clinton and Bush had to move to the center.

The privatization went far beyond Chirac's electoral promises. Initially, the 1986 law "anticipated de-nationalizing sixty-five national enterprises . . . over the next five years." But Chirac managed to privatize nearly half of the companies almost immediately; "if this pace had been kept up," Schmidt writes, "the government could have accomplished in three years what it had anticipated completing in five." According to Schmidt only the 1987 stock market crash slowed the tide of privatization. But the newly freed companies thrived. The number of shareholders in the French population grew more than five fold, from one and a half million to eight million and the economy's enormous growth allowed Chirac to pass tax cuts for corporations and individuals alike—a surefire recipe for political approval.

Why did Mitterrand let Chirac topple his socialist house of cards? Was he crazy? Crazy like a fox. By giving him his head, Mitterrand robbed Chirac of his key issue. Just as Bob Dole had nothing to say once welfare was reformed, crime reduced, and the budget deficit cut by the Clinton administration, so Chirac found himself without a platform after privatization. What, exactly, did he plan to do in the rest of his term now that he had accomplished all he had wanted to do in the first place?

Mitterrand's decision to let Chirac have his government and pass his program paid off handsomely, as the momentum of the Right evaporated. For his part, Chirac could have used the frustrations of failure to whet his appetite for power—but he couldn't deal with success. Without a central issue to guide him, Chirac was lost. After the decisive and bold success in its opening months, his premiership seemed to bounce from crisis to crisis, buffeted by events without any coherent agenda.

In his memoir of the Ford administration, *Years of Upheaval*, Henry Kissinger defines brilliantly the requisites for political leadership. "The statesman's duty is to bridge the gap between his nation's experience and his vision," he writes. "If he gets too far ahead of his people he will lose his mandate; if he confines himself to the conventional he will lose control over events." If Mitterrand had gone "too far ahead of his people" in his rapid nationalizations, Chirac now became irrelevant. With no theme and no compelling issues to define his government, Chirac found himself the victim of circumstance. When an Iranian gunboat attacked a French ship and then killed a French citizen during a hijack attempt in Geneva, Chirac broke off diplomatic relations with Iran. With French hostages in the captivity of Iranian sympathizers in Lebanon, Chirac, like Carter and Reagan, found himself mired in conflict with Tehran. Badgered by his party's right wing's desire to "keep France French," Chirac sought to curb immigration; but his policies went too far, and before long he was confronted with thirty thousand angry demonstrators in the streets of Paris protesting his efforts to limit the right of immigrants to become French citizens. As he drifted helplessly, battered by events, Chirac even began creating fissures in his own base, as his minister of culture opposed his plan to increase state censorship of the media and the arts.

As Chirac struggled to regain his footing, François Mitterrand concentrated on an area that was more likely to shore up his popularity: foreign policy. Focusing on French defense policy and disarmament policies, Mitterrand simultaneously worked to bolster relations with the Soviet Union and with Franco's Spain. "He hammered hard at the need to build a strong Europe and to improve the economic situation in France . . . [Mitterrand] announced in a television appearance that he desired 'a Europe endowed with a central political power' which decides 'the means of its security.' Shortly thereafter, Mitterrand met with Chancellor Kohl . . . [and] the two leaders . . . later agreed to create a . . . joint Franco-German military unit."

While Mitterrand basked in his foreign policy triumphs, Chirac was bedeviled by persistently high unemployment at home. But as Mitterrand never tired of implying, that was Chirac's problem, not his. After all, hadn't he given his rival a blank check to implement his economic policies? If they were flawed, it could hardly be blamed on Mitterrand the statesman, could it?

Just as triangulation allowed Clinton to embrace a values-based agenda, rising above party to speak for the common needs of American families, so Mitterrand was able to use the strategy to morph from a partisan and ideologue into the leader of a nation. "The power-sharing arrangement between 1986 and 1988," Northcutt

observes, "provided him with a unique opportunity to modify signifi-
cantly his public persona. He acted as a president who was no longer
a partisan but an umpire for the entire nation."

And there were other similarities. In 1996, Clinton consciously
remade his image, transforming himself from hamburger-munching
everyman to a dignified, dark-suited leader for the nation—at least
until Monica came along. Similarly, "cohabitation enabled Mitterrand to rebuild his popularity with the electorate," Northcutt argues.
"He became father of the nation, the impartial umpire, the architect
of industrial modernization, the defender of continuity." And he
cleverly crafted an image for himself as a unifying force determined
to protect the Constitution above and beyond political interests.

Mitterrand emphasized his newfound confidence by keeping
France guessing about whether he would run for reelection. As
Chirac fumbled and bumbled, Mitterrand looked every inch the
surefooted leader of France in the international arena, while avoiding political squabbling at home.

His popularity, which had fallen from 48 to 33 percent from 1981
to 1985, soared to 56 percent by 1988. In a survey conducted on the
sixth anniversary of his 1981 election, 58 percent called his election
"a good thing for France." His rating as president, according to the
same poll, was second only to de Gaulle's.

As the election loomed, Mitterrand led a relaxed campaign, content to let Chirac dig his own grave. He "had no laundry list of new
proposals to submit to the French," Julius Friend notes. "He ran as a
candidate in whom the people could have confidence. . . . Although
he did not renounce his past as a Socialist, Mitterrand laid his major
emphasis on reunifying the French and imbuing them with a sober
determination to face the economic challenges of the 1990s . . .
building a strong France in a more unified Europe."

The political result was a triumph. France's traditional Left, who
still supported him, were joined by a share of moderate to right-wing
voters who chose him over the directionless Chirac. And only two
years after his 1986 debacle, Mitterrand swept to victory, defeating
Chirac by fourteen points.

Mitterrand won, in part, because he changed the definition of his
job. He was able to rise to a position of prestige, even as he surrendered
day-to-day power to his prime minister. By occupying himself with the

more exalted and leaving the mundane to his rival, he was able to preserve his authority and enhance his popularity. It's a fascinating lesson for anyone who dwells in corporate or other organized structures: Absolute power can lead to absolute self-destruction—and there are times when the best thing to do is step aside and let the other fellow screw up.

Even as they used triangulation to their advantage, though, it's important to note that George W. Bush, Bill Clinton, and François Mitterrand each retained the trust of his political base. Neither truly ever repudiated his core constituency. Each simply found a way to defuse his political opponents by defusing their issues, relieving the frustrations that had spurred on their candidacies.

But triangulation is by no means a foolproof enterprise. Witness the case of Nelson Rockefeller, the former New York governor and perennial presidential candidate, who found that moving to the left meant moving out of the Republican mainstream and into the political wilderness.

NELSON ROCKEFELLER CRASHES AS HE FALLS BETWEEN THE PARTIES

It would seem that unlimited funds, national name recognition, sixteen years as a popular governor of the nation's then most populous state, an affable and attractive personality, the best staff money could buy, and relatively favorable media coverage would be enough to elect a man president—particularly if he just kept running for it. But although Nelson Aldrich Rockefeller, to the manor born, made the attempt in 1960, 1964, and 1968, he was never able to secure the nomination of his own party. Even when he was finally appointed to the vice presidency by Gerald Ford after the forced departures of Richard Nixon and Spiro Agnew, he lasted only two years and was dumped by his party at its next convention.

What was Rockefeller's sin? He triangulated. It was his willingness to straddle ideological divisions, the very trait that served Bush, Clinton, and Mitterrand so well, that destroyed Rockefeller's political career. How did he get it wrong? What can one learn from his demise to understand the strengths and limitations of triangulation as a political strategy? And what does his failure teach those who would emulate the idea?

By today's standards, Rockefeller was very, very far from your everyday Republican. But in the 1950s and early 1960s, the Goldwater-Reagan wing was still a distinct minority within the

Republican Party. A heritage of liberalism flowed through the GOP, as governors like Bill Scranton of Pennsylvania and Frank Sargeant of Massachusetts echoed Rockefeller's liberal line. Senator Hugh Scott (R-Pa.), a liberal, commanded the Republican forces in the Senate, and GOP senators like Jacob Javits (R-N.Y.), Mac Mathias (R-Md.), and Clifford Case (R-N.J.) routinely voted with liberals like Hubert Humphrey (D-Minn.) on civil rights and other unorthodox issues for Republicans. The GOP had begun moving leftward, rejecting the conservatism it had espoused in the 1920s and 1930s, after years of bludgeoning defeat at the hands of Franklin Delano Roosevelt and Harry Truman, who had beaten them in five consecutive presidential elections—the longest losing streak in U.S. history. Its presidential nominees in the 1940s and 1950s—Wendell Wilkie, Thomas E. Dewey, and Dwight D. Eisenhower—were all moderates and internationalists committed to civil rights and to most social programs. They scurried to the center to blur the differences between the parties.

Rockefeller was that kind of Republican.

Back then, the party's geographic center was in the northeast: New York, and Wall Street in particular, ruled the Republican establishment. And so, when Nelson Rockefeller contemplated launching his political career, he set his sights on the epicenter of Democratic power—the governorship of New York. It was from the Albany statehouse that Franklin Roosevelt had run for president, and from there that Democratic governors Al Smith and later Herbert Lehman forged the liberal identity of their party. The incumbent governor Rockefeller was to challenge, fellow megamillionaire and Hudson River neighbor Averell Harriman, had himself used the position to seek the Democratic presidential nomination in 1956.

Rockefeller counted on triangulation to take New York away from the Democrats. He would steal their issues, their constituency, and even their rhetoric.

Nelson Rockefeller was born a liberal. Despite the robber-baron reputation of his plundering grandfather, John D. Rockefeller, much of the family had embraced progressive causes for decades. During the Civil War, before they got rich, the Rockefellers had helped slaves escape. As Theodore H. White recounted in *The Making of the President 1964,* Nelson's "great-grandfather Spelman had run a station of the Underground Railroad in Ohio . . . to help runaway slaves get to Canada; his family had endowed the first Negro women's college (Spelman College) in 1876 . . . In all, they had given . . . $80 million to Negro causes and institutions." It was a legacy Nelson Rockefeller enthusiastically embraced: "I am proud of the contributions that my family and I have given to Negro education," he declared in 1958, "and to such great Negro organizations as the Urban League and the NAACP."

Running on a "liberal humanitarian program," Rockefeller sought the governorship of New York in 1958. Sounding as liberal as Harriman, he imbued his campaign with a recognition that "the nation's richest state was harassed by many problems, including race conflict, a housing shortage, metropolitan decay . . . and an increase in crime and juvenile delinquency." His triangulation paid off; he defeated Harriman by more than 500,000 votes.

In his inaugural address, Rockefeller spoke of the need to improve the democratic process. "We cannot speak of the equality of men and nations," he said, "unless we hold high the banner of social equality in our own communities. . . . We cannot be impressively concerned with the needs of impoverished peoples in distant lands if our own citizens are left in want."

Determined to bring black voters back to the party of Lincoln, Governor Rockefeller poured money and staff into both the state attorney general's Civil Rights Bureau and the State Commission Against Discrimination. Nobody could get to the left of Rockefeller on civil rights. He even backed school busing to achieve racial integration—a right-wing heresy. In 1972, Governor Rockefeller vetoed an antibusing bill that would have squelched nascent attempts to correct racial imbalance in schools.

Not content with stealing New York's blacks from the Democratic camp, Rockefeller anticipated George W. Bush's outreach to Hispanics by forty years as he courted New York's growing population of Puerto Ricans. He went to great lengths to create a bond with the state's Latino community, even attending Berlitz language classes for weeks to become fluent in Spanish.

As governor, Rockefeller was a nightmare to conservatives, stealing the Democrats' issues one by one. If you were a liberal, it made no sense to vote against him.

He increased welfare spending in New York from $400 million to $4 billion, and raised $277 million in taxes to pay for it. More than any other elected official in America, Rockefeller pioneered women's right to choose regarding abortion. In 1970, years before the *Roe v. Wade* decision, he took on the powerful Catholic Archdiocese of New York and its redoubtable spokesman Cardinal Cooke by urging the state legislature to approve the legalization of abortion. For the rest of Rockefeller's political career he would be labeled a "murderer" by many right-to-life activists.

Rockefeller jammed through the legalization measure by 76–73 in the state Assembly and by 31–26 in the Senate. Two years later Rockefeller vetoed a bill to repeal the legalization, saying "I can see no justification now for repealing this reform and thus condemning hundreds of thousands of women to the dark age once again."

Rockefeller's litany of Republican treachery also included opposition to the death penalty. While he signed a bill in 1965 permitting it for the murder of police officers, he refused to embrace legislation for more generally applicable capital punishment. When this limited statute was declared unconstitutional in 1973, Rockefeller refused to back a new capital punishment law.

On foreign policy questions, Rockefeller was a determined internationalist at a time when isolationism was far from dead in the Republican Party. At the end of World War II, he had played a key role in the decision of the Rockefeller family to donate the land to the United Nations on which its headquarters now stands. He was particularly focused on hemispheric issues and worked throughout the 1940s to increase American consciousness of the problems in Latin America.

The Rockefeller formula seemed to work wonders. Stealing the Democrats' issues led to Rockefeller's reelection in 1962, 1966, and 1970, as he served a record four terms as governor. Democrats were helpless against him. Anything they supported, Rockefeller got there first and advocated, funded, or enacted it.

But Rockefeller had a problem. He wanted to be president of the United States, not just governor of New York. And he would learn, bitterly, that what worked in one of America's most liberal states wouldn't fly in the rest of the nation.

By rolling out the red carpet and welcoming liberals into his party,

Rockefeller made a fatal mistake: He forgot to tack down its base. As he reached out to the progressives, his party's conservatives rolled up the carpet behind him, leaving him isolated.

A successful triangulator like Bush may have strayed from orthodoxy on education and poverty, but he remained on the reservation over abortion and taxes. Clinton moved to the center on issues like welfare reform and deficit reduction, but never betrayed the Left on affirmative action, education, the environment, or Medicare. Rockefeller, in contrast, seemed all too eager to disregard every element of GOP orthodoxy. Rockefeller moved left on *everything*.

Any leader who seeks to change his organization should heed the example of Nelson Rockefeller: The more you change, the more you must leave the same. The things you refuse to change help you preserve the loyalty of your friends, who in turn allow you the leeway to change the things you must.

How could Rockefeller back abortion, oppose capital punishment, propose higher taxes, support internationalism, fight for school busing and integration and still consider himself a Republican?

That's what Republicans were starting to wonder.

Meanwhile the progressive New York wing of the party was losing power as Eisenhower left the White House in 1960. The right wing began demanding to be heard in party councils. Their challenge was ideological, but deeply rooted in differences of geography and economic class. America's population was shifting away from the Northeast and to the West and the South. Main Street challenged Wall Street for dominance in the Republican Party. Small businessmen and religious conservatives began to chafe at the party's big business and the Eastern-banker rulers. Nelson Rockefeller, by reason of class, residence, and ideology, was their bête noire.

The clash came in 1964 when, as the representative of everything the Eastern Republican establishment stood for, Nelson Rockefeller ran against Arizona senator Barry Goldwater for the Republican nomination for president.

Goldwater tapped into a vein of the cold war paranoia that riddled some segments of the country. One supporter wrote that "New York, headquarters of the Eastern Establishment, is nearer socialist and communist influence than any of the other 49 states." To these

Republican voters, it seemed impossible that Rockefeller, who ran the People's Republic of New York, could ever lead their party.

As if his liberal positions were not enough to inflame the Republican Right, in 1962 Rockefeller chose to divorce Mary Todhunter Clark, his wife of thirty-two years and the mother of his five children, and marry thirty-seven-year-old Margaretta "Happy" Fitler Murphy only fourteen months later. The move came less than a year before his first presidential campaign began, and the timing could not have been worse. To religious conservatives in the West and South and Catholics in the Northeast, the divorce loomed large as a political negative. America's innocence in such matters had not yet come to an end. Only one presidential candidate had ever been divorced—Adlai Stevenson—and he had lost two consecutive bids for the White House.

Theodore S. White underscored the importance of the divorce in *The Making of the President 1964.* "One can no more discuss the Republican politics of 1964 without dealing with Nelson Rockefeller's divorce and remarriage than one can discuss English Constitutional development without touching on the stormy marriage of Henry II and Eleanor of Aquitaine."

When Rockefeller remarried, White reported, the Chicago Young Adults for Rockefeller "disbanded in anger. Ex-senator Prescott Bush of Connecticut, a longtime friend of Rockefeller's [and patriarch of the future presidential Bush clan] branded him 'a destroyer of American homes.'" Rockefeller took flak not only for his divorce and remarriage, but for his new wife's conduct in leaving her family to marry him. White notes that the feeling among American women was, "I ain't going to vote a woman into the White House who left her children."

The 1964 California Republican primary—the major confrontation between Rockefeller and Goldwater—began to focus far more on Rockefeller's marital history than on the ideological issues that divided the two candidates. The birth of Nelson Rockefeller, Jr., Rockefeller's son with his new wife, on the eve of the Republican primary in California only served to remind voters of the scandal.

In the month before the primary, the polls began to reveal how severe a toll the divorce issue had taken. Goldwater, who had trailed Rockefeller among Republican voters by seventeen points, now led him by five. Goldwater continued to pound away, and ultimately knocked Rockefeller out of the race by capturing all of California's delegates.

Why did Rockefeller risk the divorce? He was arrogant, and in the last analysis, a political dilettante. He was willing to spend from his fortune and devote his energy to getting elected to the presidency. But he was not prepared to subject himself to the discipline the political process demands. At a core level, he couldn't be bothered to let the electorate impose its views upon him, even for the year or two he needed to get elected.

Like Bill Clinton, Rockefeller was not willing to discipline his private conduct to achieve success in elective office.

Bruised and damaged in defeat, Rockefeller encouraged Pennsylvania's moderate Republican governor, William Scranton, to take up the anti-Goldwater banner. Scranton, who labeled Goldwater "dangerous," brought the ex-Rockefeller supporters to his side.

Not to be deterred, Rockefeller decided to "deliver a minority report to counter the harsh Goldwater platform plank condoning political extremism," as Joseph Persico recalled.

The New York governor knew, of course, that he could no longer stop the Goldwater bandwagon. But he hoped to use his opposition to the senator's right-wing agenda to capture the hearts of his party's rank and file. But the Republican Party had changed before his eyes; he just hadn't grasped it until it was too late.

As Rockefeller rose to address the 1964 convention, catcalls cascaded down upon him. As the *New York Times* reported, the "outburst" and "boos" Rockefeller encountered moved him to snap at the delegates: "Some of you don't like to hear it, ladies and gentlemen, but it's the truth." He was interrupted continually, but Rockefeller left the podium with a smile, to a warm greeting from New York congressman (and future mayor) John V. Lindsay of New York City, a fellow moderate Quixote within the GOP ranks. (The anti-extremism plank was defeated.)

Why had triangulation failed Rockefeller? How did he stumble where Bush, Clinton, and Mitterrand succeeded in moving their parties to the center?

Rockefeller had misjudged the feelings of his own party. George W. Bush understood that Republicans were uncomfortable seeing themselves as hard-hearted on poverty, and realized that the young parents in his party wanted to do more to improve their children's schools. Clinton knew that fear of crime and frustration with welfare costs reached deep into the Democratic base, and that he could move to the center on these issues without losing them. Mitterrand saw that his own party recognized the practical failure of socialism, and perceived that he could shelve the work of a lifetime without losing his base.

But Rockefeller failed to understand how the fiscal conservative small-town businessmen of the South and the West had taken over his party. No longer were its positions shaped by Wall Street tycoons and their heirs, who were so financially secure that they could afford a little liberalism. The very rich had ceded their power in the GOP to the working middle class. For them, high taxes, growing welfare, and school busing were threats to their fragile grip on the good life in the suburbs.

Unlike Bush, Clinton, and Mitterrand, Rockefeller discarded all the ideological commonality between himself and the bulk of his party. He strayed on everything, and was loyal to his base on nothing. It was the difference between triangulation and simply switching sides. Rockefeller himself admitted as much: When an Argentine journalist asked "Senor Rockefeller, why were you never elected your country's president?" Rockefeller responded, "I was in the wrong party." Rockefeller even reported that President Harry S. Truman "tried to get me to change [parties] . . . when I was working for him" in 1945. Rockefeller refused, explaining "I would rather try to pull a group forward than hold a group back."

But politics was far from over for Nelson Rockefeller, who got a second chance after his 1964 defeat. Despite his loss, Rockefeller felt vindicated by Lyndon Johnson's landslide victory over Goldwater in 1964. To many Americans—and likely many Republicans—Rockefeller now seemed a voice of sanity overwhelmed by the unseemly but limited band of extremists who had hijacked the party and sacrificed the election. Many assumed that Rockefeller would emerge in 1968 as the likely GOP nominee now that the Republican Party's flirtation with conservatism had ended in such disaster.

But a new dynamic started immediately after the 1964 election, as President Lyndon B. Johnson ignored his campaign promise not to "send American boys to do what Asian boys ought to be doing for themselves" and committed the United States to a full-scale war in Vietnam. With hundreds of thousands drafted and tens of thousands killed, political forces in America seemed to shift like tectonic plates in an electoral earthquake as the nation became mired in the Vietnam War. Johnson came to personify the mindless escalation of the war, as half a million American soldiers trudged into a battle of constant attrition with a tenacious foe. The peace movement gained in strength with each month.

As the nation drifted to the left, Rockefeller missed his shot once again. He was like the French Army General Staff of whom it was said that they entered each new war "perfectly prepared to fight the last one." It was a tragic and instructive lesson in battle readiness: It is far easier to content oneself mulling the lessons of the past than to anticipate and prepare for the changes the future holds. Having missed the migration of the GOP from Wall Street to Main Street, Rockefeller blinked—and missed the nation's move to the left over Vietnam.

The winds of 1964 had shifted too quickly. Rockefeller had been too liberal to capture his party's nomination in that year; if he had stayed in the center then he might well have won. But by 1968, when Rockefeller's natural left-leaning tendencies might have won him the support of students demonstrating against the war, Rockefeller went the wrong way, backing the war as the peace movement came to see him as its apologist.

Anticommunism had always been central to Rockefeller's ideology, as his family's capitalist background would suggest. Indeed, he had the distinction of being one of very few Americans whose ancestors had been attacked by name by Karl Marx and V.I. Lenin. For him, anticommunism was a family grudge.

Rockefeller's well-known liberalism motivated him to show his anticommunist credentials, so as not to fall too far to the left in the public's mind. Like Johnson, he shared the establishment consensus reflected in John F. Kennedy's inaugural address that "we shall pay any price, bear any burden, meet any hardship, support any friend, oppose any foe to assure the survival and the success of liberty."

In his biography *The Imperial Rockefeller,* Persico calls Rockefeller a "hawk" who saw "communism [as] . . . a threat to the system that had blessed his family and his country. The Vietnam War was being fought to stop communism . . . therefore, [to him] it was a just war."

As millions of young people protested the war in the streets, they searched in vain for a candidate to challenge it in the elections of 1968. But the Democrats were slow to take on Johnson. Robert F. Kennedy, the former attorney general and brother of the slain hero-president, was the obvious choice, but at first he demurred despite his outspoken opposition to the war his brother had started. Among Republicans, Richard M. Nixon, a lifelong anticommunist zealot, was the leading candidate.

The Left turned its lonely gaze to Rockefeller. But the Republican liberal, while as eager as ever to run, showed no eagerness to flank Johnson on the left over the war. In fact, he seemed more to criticize Johnson for failing to prosecute it successfully. Where the Left sought an advocate for peace, they found in Rockefeller only support for a more effective war. At the crucial moment, Rockefeller so profoundly disappointed the aspirations of his liberal constituency that he was left behind by the march of history.

Instead, others raised the antiwar banner. Senator Eugene McCarthy, a liberal Democrat from Minnesota, emerged to challenge Johnson in the Democratic primaries, while Rockefeller battled Nixon for the Republican nomination. After McCarthy did very well in the New Hampshire primary, RFK reconsidered his earlier passivity and entered the race against Johnson. The president withdrew, McCarthy faded, Kennedy died, and the Vice President, Hubert Humphrey, won the Democratic nomination.

Having lost his bid to transform the Republican Party in 1964, Rockefeller had attempted to realign himself with its orthodox base as he campaigned for the nomination in 1968. But the GOP did not trust him. Rockefeller could have reshaped American politics by firmly opposing the war: in the general election America was left with three candidates—Humphrey, Nixon, and segregationist governor George Wallace of Alabama (running as an Independent)—who were all ardent hawks. Had Rockefeller run as an independent on an antiwar platform in 1968, while three candidates split the prowar vote, he might well have been elected. But at the crucial moment he lost his nerve; when the nation moved to the left he moved right, and Richard Nixon won the election.

By the 1970s Rockefeller should have known that his future in the Republican Party was limited. Even as he attempted to move to the right on issues of crime and drugs in the 1970s, he could never recapture the hearts of the GOP.

His governorship drifted further and further to the right in a vain courtship of the party's conservatives, as he passed drug laws that mandated harsh sentences for relatively minor offenses such as the possession of marijuana. He raided Attica Prison when it was taken over by inmates protesting terrible conditions. After they took dozens of guards as hostages, mediators began delicate negotiations. Frustrated with their pace, angered by the inmates' rhetoric, and determined to show his toughness to a watching nation, Rockefeller ordered an all-out assault that recaptured the prison, but led to many inmate and hostage deaths.

When Nixon was forced from office in 1974, the new president, Gerald Ford, reached out to nominate Rockefeller as vice president. Attacked from both the left and the right, Rockefeller was confirmed in the House of Representatives by a margin of 287 to 128. At the time Rockefeller reflected, "There were about a hundred and twenty conservatives and liberals, so identified. I got the opposition of both. Those in the middle voted for me. And that is about where I stand."

But the Right would never trust Rockefeller, despite his recent conservatism, and refused to accept him on the 1976 ticket with Ford. As Reagan challenged the Republican president for the party nomination and won almost half the delegates, Ford jettisoned Rockefeller as a political liability in favor of the conservative Senator Robert Dole of Kansas.

In what biographer Persico calls "a ritual act of political hara kiri," Rockefeller was asked to place Dole's name in nomination. His agreement to do so marked the end of his strange and thwarted political life.

The lesson of Nelson Rockefeller is: watch your base. When you move to open new markets or seek new votes, be careful to retain enough commonality with your traditional supporters to keep them in the fold. Rockefeller didn't, and he learned that he who doesn't watch his base had better watch his back.

Rejected by his own party, unwilling to reach out to the opposition or form his own, Nelson Rockefeller watched his political career fall between two chairs, lost forever.

STRATEGY THREE

DIVIDE AND CONQUER

Dividing your adversaries, splitting them asunder, and pitting them against one another is the oldest of military and political stratagems. The quintessential Roman strategy for subjugating the world, the divide-and-conquer gambit has a particular relevance to the world of electoral politics.

Despite the United States' vaunted two-party system, one-third of our presidents in the past century have been elected with less than a majority of the popular vote, as third and fourth parties have siphoned away votes from the major party contenders. Indeed, our last three presidential elections were decided by less than a simple majority (and our most recent by less than a plurality!).

How do you go about dividing your enemies in order to conquer them? What can induce an adversary to stray down a self-destructive course?

Although Abraham Lincoln and Richard Nixon have little more in common than that they were Republican presidents who were each elected by less than a majority vote, each man divided and conquered. But they used opposite strategies to trigger the split in their opponents' ranks. Lincoln took a strong position on the slavery issue as he prepared to run for president, in part, to split his Democratic opponents and weaken their ranks. Nixon, on the other hand, took no position on the war in Vietnam as he ran in 1968, ducking the issue so that the division already rending the Democratic party would carry on uninterrupted, leaving him an opening to step into the presidency.

A third case—that of Dewey against Truman in 1948—reveals that that dividing doesn't always mean conquering. When the Left deserted Truman over his stance against communism—and the Right split over his attacks on racism—the fissures empowered him. He used defections on the left and the right to help highlight his independence and dedication to principle. The creative use of dissent: it's an intriguing weapon in the pursuit of power, one that has served three politicians as diverse as Honest Abe, Tricky Dick, and the Man from Independence.

LINCOLN SPLITS THE DEMOCRATS OVER SLAVERY AND GETS ELECTED

The greatest president in American history almost wasn't elected. A radical on the key issue of slavery, Abraham Lincoln was the choice of only 40 percent of the voters. Six in ten voted against him. It was only because the Democratic Party fractured into northern and southern wings that Lincoln won.

But the Democratic breakup wasn't accidental. It was the calculated result of a strategic masterstroke envisioned, planned, and implemented by Abraham Lincoln—not only the greatest president of all time, but also one of our craftier politicians.

On April 23, 1860, the Democratic Party met to choose its nominee for president. Having controlled the White House for the past eight years and twenty-four of the previous thirty-two, the chieftains of the party realized that if they could only hold their ranks together, another victory might well be within their grasp. But it was not to be. Two days before the convention opened, the delegations of eight southern states, Georgia, Alabama, Mississippi, Louisiana, Florida, South Carolina, Arkansas, and Texas, vowed to insist on a pro-slavery platform and threatened to leave their party unless their position was accommodated.

The Democratic nominee, Senator Stephen A. Douglas of Illinois, embraced a more moderate view: that the residents of each state or territory should decide whether to allow or forbid slavery. And when the Democratic convention adopted Douglas's recommendations, the southerners bolted, walking out on April 30, 1860. Assembling in Baltimore on June 18, they nominated the sitting vice president, Kentucky's pro-slavery John C. Breckinridge, as their candidate for president.

Once the Democrats split, the Republican nominee, Abraham Lincoln of Illinois, believed that the tide of the forthcoming election had turned inexorably in his party's direction. Lincoln "had little doubt that the Republicans would win the presidential election," biographer David Herbert Donald notes, "if the discordant and rival elements that composed the party could work together."

Lincoln's political forecast proved accurate. Running against a split Democratic Party, on election day he carried every northern state (except for New Jersey, which went for Douglas), while Breckinridge beat Douglas in every southern state. Though he garnered just 40 percent of the vote, Lincoln was elected president, solely due to the division in the Democratic ranks.

How did Lincoln divide in order to conquer?

In the tumultuous period leading up to the Civil War, the Democratic Party had ruled the nation—and held itself together—by straddling the slavery issue. Democrat presidents Franklin Pierce in 1852 and James Buchanan in 1856 had won on the strength of a simple formula: they came from the North, but thought like the South. Pierce, from New Hampshire, and Buchanan, from Pennsylvania, were unabashed in their southern sympathies. Nicknamed "doughfaces" for their weak and mushy politics, they managed to maintain party and national unity through years of government by ambiguity and obfuscation.

Outvoted in the House of Representatives by the more populous North, the South pinned its hopes for preserving slavery on the Senate, where it banked on an equal number of slave and free states. But as migration opened up new territories in the west, the balance in the Senate became more and more precarious.

Southern fears were calmed by two great deals brokered by Henry Clay—the Missouri Compromise and the Compromise of 1850— which together sketched a line along the 36° 30' N Lat parallel, along the southern border of Missouri. When this Missouri Compromise line was eventually extended all the way to the Pacific in 1850, it separated an entire continent into a free north and a slaveholding south.

But this balance was upset in the mid-1850s, when Lincoln's future Senate and presidential opponent, Stephen Douglas, the leading Democratic Party statesman of the time, began urging that the slavery issue be decided by voters in each territory as it became a new state. This idea, which became known as "popular sovereignty," or later states' rights, would set the stage for civil war and lead directly to the division in the Democratic Party that elected Lincoln.

First embodied in the Kansas-Nebraska Act passed by Congress in 1854, the theory of popular sovereignty caused a mini-exodus as northern and southern settlers alike packed up and thronged into the Kansas territory, intent on leveraging their votes to decide the slavery issue. When it became clear that the northerners had won the migration race and would win any fair election, the southerners resorted to violence and fraud. Excluding northerners from voting, they jammed through the Lecompton Constitution, which made slavery the law of the territory.

Douglas, angered by their tactics, condemned the Lecompton Constitution, a position that strongly alienated some of his former southern supporters. Georgia's former governor, Herschel V. Johnson, warned in May 1858 that if Douglas should be president "it will be in despite of [sic] the South."

Meanwhile, outraged at the slaveowners' tactics, antislavery forces matched southern violence and intimidation with their own. Abolitionists, led by John Brown, attacked southern settlers in Kansas, and Douglas's popular-sovereignty doctrine blew up in his face as the North battled the South by proxy in Kansas. "The Lecompton crisis left a legacy of party disruption and sectional hatred from which the nation would not recover," historian Robert Johannsen writes, "and Douglas was the chief inheritor of that legacy."

Lincoln, for his part, opposed both Clay's compromises and Douglas's popular sovereignty plan, demanding instead that any new state must be declared free. He believed that the Constitution made it impossible to emancipate slaves without an amendment. But he felt it was within the power of Congress to require that a state newly admitted to the Union must ban slavery as a condition of its entry—and he hoped that the hated institution would wither away in consequence, as more and more free states joined the growing nation.

The South, sensing that it was losing the national battle, turned things completely on their head when Chief Justice—and slaveowner—Roger Taney delivered the infamous Dred Scott decision. In a 7–2 split, the Supreme Court ruled that Dred Scott, a slave who had fled from the South and lived as a free man in the North, could be legally recaptured by his former owner and returned to slavery in the South. The ruling struck down the Missouri Compromise line as unconstitutional. As for slaves, they were property, nothing more—and the Constitution obliged each state to honor the property of a citizen of another state. No state or territory could stop anyone from bringing his slave into their jurisdiction and keeping him enslaved. To require that a man free his slave just because he moved to a free state, the court implied, would be as unjust as to ask him to give up furniture or animals that might have moved with him.

The Dred Scott decision was the final straw for northern abolitionists. Until it was handed down, they could content their consciences with the hope that slavery would eventually die out on its own as more and more nonslave states came into the Union. But the court decision raised the inescapable prospect of a permanent slave system, not just in the South but in every state of the nation.

The ruling ripped the Democratic Party apart. While President James Buchanan backed the decision—indeed, he had secretly and unethically lobbied Taney to rule as he did—and moved to force slavery on Kansas, Senator Stephen Douglas, the other leading Democrat of the era, demanded that popular sovereignty be honored.

At a fateful White House meeting, President Buchanan demanded that Douglas back him in supporting the Dred Scott decision. When the Illinois Senator demurred—possibly out of concern for appeasing antislavery sentiment in his home state—Buchanan declared political war on Douglas, undermining him in his own state. Buchanan's animosity was so great that he even removed the postmaster of Cook County and replaced him with a long-time rival of Douglas.

It was against this background that the Illinois lawyer and former Whig Congressman Abraham Lincoln decided to challenge the redoubtable Douglas for the Illinois seat in the United States Senate in the election of 1858, using slavery as his major issue.

Lincoln may have hoped to profit from the Buchanan–Douglas feud, but he certainly had larger plans in mind. His goal was to defeat Douglas, not just for the Senate but for the presidency as well.

Lincoln's plan was to up the ante on the slavery issue in Illinois, exploiting the northern state's contempt for the practice to force Douglas to take stronger and stronger positions against slavery. By condemning human bondage in the harshest of terms, Lincoln was, in effect, daring Douglas to follow suit. If Douglas took the bait, he'd likely be reelected to the Senate by liberal Illinois, but would lose all hope of southern backing, which would be essential in his presidential bid two years later.

Asked about what makes a successful politician, Lincoln observed that what was required was "to be able to raise a cause which will produce an effect, and then fight against the effect." Lincoln's cause was his denunciation of slavery. His effect was Douglas's move to the left on the issue. He would then, in 1860, fight against the effect, which was a split in the Democratic Party.

A seasoned trial lawyer, Lincoln was as skilled as any chess master at figuring out political tactics two or three moves ahead, as he demonstrated in his debates with Douglas.

When Lincoln challenged Douglas to debate, the Democrat unwisely accepted, opening the way for Lincoln to spring his trap on him. His opening gambit was to lure Douglas into his trap by staking out a strong moral antislavery position. In speeches in Chicago and Wisconsin before the debates began, Lincoln spoke of his "hatred to the institution of slavery" and committed himself to its "ultimate extinction."

As Lincoln confronted Senator Douglas in the actual debates, he drew sharp distinctions between his positions on the slavery issue and that of his opponent. Sometimes called a "Black Republican," Lincoln looked at slavery "as a wrong—a moral, social, and political wrong."

Addressing those who held viewpoints different than his own, Lincoln said, "If there be a man amongst us who does not think the institution of slavery is wrong . . . he is misplaced and ought not to be with us [in the Republican Party]."

Lincoln portrayed Douglas as two-faced, a man trying to have it both ways—to please both those who favored slavery and its opponents. "He has never said . . . that the people of the Territories can exclude slavery," Lincoln charged. Lincoln tried to show that Douglas's view of popular sovereignty was tantamount to trying to play both sides of the fence, that it was just another way to allow slavery to continue while seeming to appeal to freedom.

Lincoln mocked popular sovereignty, saying that while "each man . . . [can] do precisely as he pleases with himself, and with all things which exclusively concern himself," that Douglas thought it was acceptable "if one man chooses to make a slave of another man, neither that man or anybody else has a right to object."

As historian William Baringer wrote in 1937, Lincoln "called upon Democrats who honestly opposed slavery to see the delusion of the Douglas teaching that 'popular sovereignty is as good a way as any to oppose slavery' to desert Douglas and give their support to the only real instrument of opposition to slavery, the Republican Party."

Lincoln persisted in his attacks, rebuking Douglas for his nonchalant view of slavery. By arguing that the laws of slavery were no different from those regulating liquor or property, Douglas seemed the moral inferior of those who opposed slavery. "This declared indifference, but as I must think, covert *real* zeal for the spread of slavery, I can not but hate. I hate it because of the monstrous injustice of slavery itself." The primary difference between the two parties, Lincoln argued, was to be found in their moral views of slavery.

Douglas was vulnerable to Lincoln's charges. While Douglas was not a slaveholder himself, he almost ostentatiously refused to condemn the institution, and took a firm stance against abolition. Anxious to run for president, Douglas knew he could not afford to lose the support of the South, essential to any Democratic candidate.

But Lincoln, by aggressively challenging him on the issue, was calling his bluff. As David Zarefsky writes, "Lincoln was frank in acknowledging that his purpose in putting the issue in this fashion was to realign the audience by winning over to his own side some who now supported Douglas." As the famous Lincoln–Douglas debates unfolded, the challenger continued to lay his trap in the Alton debate of October 15, 1858. "Each man currently numbered among his supporters some who thought slavery to be at its root an evil. But, 'whenever we can get rid of the fog which obscures the real question,' this element can be brought to the Republican side. That shift in allegiance, Lincoln thought, would be sufficient to set slavery on the course of ultimate extinction." Conscious that he was alienating the South, Douglas nevertheless felt obliged to answer Lincoln's challenge on the slavery issue. Illinois, after all, was a northern state; slavery had been outlawed there for more than half a century. The fact that Lincoln was in Douglas's face every few days during their debates made it nearly impossible for the Democrat not to respond to the Republican's barbs over the slavery issue.

Lincoln denied Douglas the option of ambiguity on the slavery question by his fierce questioning during their second debate at Freeport, Illinois, on August 27, 1858. "Can the people of the territories exclude slavery?" Lincoln asked his opponent.

Douglas answered yes.

Then Lincoln asked his opponent, "Can the people of a United States territory, in any lawful way, against the wishes of any citizen of the United States, exclude slavery from its limits prior to the formation of a state constitution?" In other words, would he follow the Dred Scott decision or not?

Douglas bit—and effectively nullified his presidential viability with his answer: "I answer emphatically . . . that in my opinion the people of a territory can, by lawful means, exclude slavery from their limits prior to the formation of a state constitution." In other words, Dred Scott didn't matter. "It matters not what way the Supreme Court may hereafter decide as to the abstract question whether slavery may or may not go into a territory under the Constitution, the people have the lawful means to introduce it or exclude it as they please. . . . Hence, no matter what the decision of the Supreme Court may be on that abstract question, still the right of the people to make a slave territory or a free territory is perfect and complete under the Nebraska Bill."

Douglas's repudiation of the Dred Scott decision may have helped him to win the Senate seat, but it cost him the White House. According to Zarefsky, in his foreword to the text of the Lincoln-Douglas debates, "Douglas' victory [in the senatorial election] was a costly one. His answer to Lincoln's second Freeport question— that the territories need not have slavery in spite of the Dred Scott decision—seemed to southern extremists to be a prodigal discarding of a hard won right. Douglas' position was not new . . . but never before had he been compelled to expound, reiterate, and elaborate it before a national audience. Since it labeled him as indifferent to the spread of slavery, rather than an advocate of the institution, the extreme proslavery leaders of the Democratic party counted him out. In 1860 they would split the party and insure Lincoln's election rather than accept Douglas as the nominee."

Carl Sandburg agreed, noting that Douglas's answer "raised a storm of opposition to him in the South and lost him blocks of northern Democratic friends who wanted to maintain connections in the South."

Lincoln had used his Senate candidacy and the debates with Douglas to pin his opponent down on the slavery issue with such clarity that he could no longer be elected president. Lincoln had lost the Senate battle, but had forced Douglas to box himself in so thoroughly that the Republican was guaranteed to win the presidential war.

Was Lincoln, in fact, thinking two or three moves ahead when he popped the question to Douglas in Freeport? Did he expect that, by pinning his rival down on blocking the spread of slavery, he could force him to choose between being Senator and being president? Who knows?

Carl Sandburg says that "Lincoln showed his [Freeport] questions to advisors beforehand; they told him to drop the main question. He answered, 'I am after larger game; the battle of 1860 is worth a hundred of this.' His guess was that Douglas's answer would split the Democratic party and make a three-cornered race for the Presidency two years later."

Another Lincoln biographer, David Herbert Donald, doubts the Sandburg account, calling it a "legend." Legendary or not, Lincoln must have known that pinning Douglas down on an issue where his rival sought refuge in ambiguity would impair Douglas's political health, and make him unacceptable either to Illinois or to the South.

Two years later, as the two men battled for the presidency, it was clear that the slavery issue still dogged the Democrat's footsteps. Douglas compounded his difficulties in a speech at Norfolk, Virginia, by saying that if Lincoln were elected, the southern states would not be justified in seceding. Would he act militarily against them if they did? The answer he gave further decreased his popularity in the south: "It is the duty of the president to enforce the laws and he would do all in his power "to aid the government in maintaining the supremacy of the laws against all resistance."

As Johannsen notes, "Douglas' 'Norfolk Doctrine' was immediately denounced in the south as counseling coercion against the states. Once again he was charged with treason, and southern radicals used his statements to bolster their arguments for secession." A Washington newspaper accused Douglas of completing "his transition to the Republican party." His Norfolk speech, it said, included some of the "boldest defiances and most pregnant warnings to the south that could be uttered."

Douglas's comments heartened the supporters of Vice President John Breckinridge, who had resolved to make his own bid for the Democratic nomination on a largely pro-South platform. While the vice president looked south for much of his support, his northern backers hoped that Douglas's strong stand against secession might generate support for Breckinridge even in the northern states, as the only way to avoid a civil war. In the end, Breckinridge and his followers bolted the Democratic Party, and he ran for president as the candidate of a separatist southern wing of the party. He carried all the southern states but was shut out in the North. It was this split that elected Lincoln president.

Lincoln's gambit had paid off—perhaps better than he'd expected. Coming from a minority party, he knew he would have to weaken the majority party in order to win; by forcing Douglas to the left on slavery, he found he had accomplished something even more dramatic—he had fractured the unity of the Democratic Party.

Squirming to escape the trap in which Lincoln had caught him, Douglas tried to appease the South by backing the platform adopted by Breckinridge which pledged that "all questions pertaining to slavery should be banished from Congress; that the people of the territories be free to make their own laws and regulations respecting slavery." But the concession came too late. By now, the southerners had taken irreversible stands against Douglas, branding him a heretic to traditional Democratic views, and demanding a candidate who would ensure the protection of slavery in the territories. Frequently they threatened party disruption if the Democratic national convention should fail to adopt a pro-South platform. Douglas had become too liberal for the South. As one Southerner implored: "If Douglas and disunion were presented to their choice . . . adopt the latter as the less evil!"

On April 30, 1860, after the Democratic Convention adopted a platform embracing Douglas's popular-sovereignty stance on the slave issue, the delegates from Alabama, Mississippi, Louisiana, South Carolina, Florida, Texas, and Arkansas withdrew. The next morning the Georgia delegation walked out. The Democratic Party had torn itself in two.

As Southern Democrats gathered in Baltimore to nominate Breckinridge for president, Douglas set about trying to salvage what votes he could in the South. But in the southern press he was now being "portrayed as Lincoln's ally"—a piquant irony, given that the very statement that gave rise to this perception occurred as the two men engaged in what would become the most celebrated debates in American history. A Mississippi newspaper warned that the South would " 'bristle with armed men' to repel the aggressions of Lincoln and Douglas and Jefferson Davis, [future president of the Confederacy], reportedly advised his [Mississippi] constituents to greet the invasion of Lincoln and Douglas with a gallows that took into account only their difference in height." As Douglas concluded his tour, an Alabama newspaper hurled the final warning. "Douglas did well to turn his course Northward. There are some portions of the South where the utterance of such sentiments might have led to the hoisting of that coat tail of his that hangs so near the ground to the limb of a tree, preceded by a short neck with grapevine attachment."

Douglas wasn't lynched, but he didn't win a single southern state either. Lincoln triumphed by exploiting the divisions he had done so much to sow in the Democratic Party. In their Freeport debate of 1858, Lincoln had divided. In the 1860 election, he conquered.

What is the strategic lesson of Lincoln's ploy? When you face an adversary who is straddling the fence and trying to have it both ways on an issue, smoke him out by questioning him directly. He won't be able to avoid giving an answer any more than Douglas was at Freeport. Don't let him hide behind ambiguity. Take a clear stand yourself and then pin him down, too.

More than a century later, Richard M. Nixon chose the opposite way of dividing his enemy. If Lincoln divided the Democrats by forcing a point of principle against slavery, Nixon divided his rivals by keeping his public stance ambiguous, allowing the American public to read into his silence whatever it wished. Concealing his real views on the war that raged in Vietnam, Nixon was able to play the Democrats off against each other to achieve the split he needed to win the presidency.

NIXON CAPITALIZES ON THE DEMOCRATIC SPLIT ON VIETNAM TO GET ELECTED

Abraham Lincoln divided the opposition through courage and forthrightness. His bold and honest stand against slavery made his opponent's attempts at evasion impossible to sustain.

But honesty is not the only way to split the enemy. Richard M. Nixon proved, in 1968, that a shameless politician can accomplish the same effect—at least in the short term—by wearing false colors into battle.

Nixon, the ultimate cold warrior and military hawk, was elected president in 1968 because he managed to convince enough Democrats that he was dovish on the Vietnam War to prevent the opposition from coalescing around his opponent, Hubert H. Humphrey. In a campaign full of ambiguity and deceit, Nixon so fudged and falsified his position on Vietnam that antiwar Democrats sat on their hands and let him win.

Nixon's flirtation with peace in Vietnam lasted no longer than the year it took him to get elected. Before he entered the race he had spent two decades as the most rabid—and visible—opponent of communist expansion. After he won he escalated the war in Vietnam and prolonged it, doubling the American death toll in the process.

Nixon did not create the Democratic division over Vietnam, any more than Lincoln made the Democrats disagree over slavery. But Nixon exploited the division and exacerbated it to get himself elected by pretending to be something he was not—a foreign-policy moderate.

In the 1940s, Nixon, more than anyone else, used anticommunism as a political issue, making his pose of moderation particularly disingenuous. Nixon had played on American fears of communism since 1946 while running for Congress against liberal Democratic incumbent Jerry Voorhis. In a time when Joseph Stalin was annexing eastern Europe to the Soviet empire, Nixon effectively painted his opponent as a communist sympathizer—and rocketed to victory.

In Congress, he heightened his reputation by serving on the House Committee on Un-American Activities. When Alger Hiss, a respected diplomat who had been one of FDR's advisors at the Yalta Conference with Stalin, appeared before the committee and denied that he was a communist, Nixon set out to prove that Hiss was lying. When he succeeded—rightly, it would emerge years later—Hiss was sentenced to five years in prison, and Nixon became a national figure.

Exploiting the Hiss conviction, Nixon ran for the Senate against Helen Gahagan Douglas, wife of the well-known film actor Melvyn Douglas. Claiming that Douglas was "pink right down to her underwear," Nixon said that "if she had her way . . . Alger Hiss would still be influencing the foreign policy of the United States." Nixon accused Douglas, Secretary of State Dean Acheson, and even President Harry Truman of harboring communist sympathies. Sensing what Nixon was up to, Helen Douglas retorted: "The BIG LIE. Hitler invented it. Stalin perfected it. Nixon uses it." But to no avail. Nixon trounced her to become California's new Senator.

When the Republicans nominated World War II hero Dwight D. Eisenhower for president in 1952, Nixon's high-profile battle against communism earned him the nomination as vice president. Continuing to play the red card, Nixon led the attack on Democratic nominee Adlai E. Stevenson. As he writes in his memoirs, "I used a phrase that caught the public's attention . . . when I charged that Stevenson was a graduate of Acheson's 'Cowardly College of Communist Containment.'" Contrasting the unimpeachable Eisenhower with Stevenson, Nixon said he would rather have a "khaki clad president than

one clothed in State Department pinks." Nixon's red-baiting helped
the Republicans win by ten points, capturing the White House for
the first time in twenty years.

Nixon continued to pound away at anticommunist themes in
1960, when he ran for president against John F. Kennedy, calling for
tough action against Cuban dictator Fidel Castro. But after an
incredibly narrow defeat in 1960 and a subsequent loss in a race for
governor of California, Nixon suddenly seemed to have hit a political
wall. Realizing that his inflammatory rhetoric was no longer likely to
gain him any political advantage, Nixon scurried toward the political
center.

Over the next several years, Richard Nixon conducted a careful
and calculated campaign to sell the American public a "new Nixon,"
no longer the nasty, negative man voters had learned not to like and
the media loved to hate. When Nixon's archenemy of old, John F.
Kennedy, was assassinated in November 1963, Nixon revisited—and
revised—the history of their relationship, telling the media that he
and Kennedy were friends after all. "I was as friendly with him as he
was with any senator on the Republican side," Nixon said. "To some
people he was President, to some a friend, to some a young man. To
me he was all that, and on top of that, a man of history."

After Johnson defeated Goldwater in 1964, Nixon grew determined
to build a new image for himself as a moderate Republican. Writing in
a monthly column in 1965, this man who would later use the "southern
strategy" of risking the alienation of blacks in order to win southern
white support, warned that "Republicans must not go prospecting for
the fool's gold of racist votes. Southern Republicans must not climb
aboard the sinking ship of racial injustice. They should let Southern
Democrats sink with it, as they have sailed with it."

And in the meantime, Nixon watched and waited for an even
more tempting opportunity: to position himself as a moderate on the
growing war in Vietnam. When President Johnson returned from a
meeting in Manila that seemed to point to another escalation in the
war, Nixon asked, "How many more American troops—in addition to
this latest 46,000—do we currently plan to send to fight in Vietnam
in 1967?" His question hung in the air, just as he had intended it
would, leaving the distinct impression that he was against the war.

Rising to the bait, Johnson lashed out at Nixon's criticism, publicly complaining that Nixon "doesn't serve his country well" by making such remarks. For once, the press actually sided with Nixon, reinforcing the suggestion that he was more liberal on Vietnam than Johnson. As Nixon confidant Earl Mazo wrote, by answering Nixon's attack "overnight Lyndon Johnson had transformed Nixon from *a* Republican leader into *the* Republican leader."

For Nixon to criticize Johnson over the Vietnam involvement was an act of incredible chutzpah. After all, it was the memory of Nixon and other Republicans hounding Truman over the loss of China to the communists that helped induce Kennedy and Johnson to intervene in Vietnam in the first place. As Johnson explained: "I knew if we let Communist aggression succeed in taking over South Vietnam, there would follow in this country an endless national debate—a mean and destructive debate—that would shatter my Presidency, kill my administration, and damage our democracy. I knew that Harry Truman and Dean Acheson had lost their effectiveness from the day that the Communists took over in China. I believed that the loss of China had played a large role in the rise of [Senator] Joe McCarthy. And I knew that all these problems, taken together, were chickenshit compared with what might happen if we lost Vietnam."

Nixon's efforts to drape himself in moderate-liberal clothes as he cast an eye toward 1968 speak volumes about Nixon's own talent for duplicity—but also about the unpredictable political environment of the mid-1960s, a time when traditional precepts and presumptions seemed to be collapsing before the naked eye. As the Democratic president waded deeper into the swamp of a no-win war against the Communists and the American troop commitment—along with casualty lists—mounted, Nixon, the anticommunist Republican standard-bearer, stayed out of sight. Determined to give Johnson enough rope to hang himself, Nixon carefully avoided being drawn into any discussion that might compromise his eventual flexibility to say what he needed to win in 1968. For six months in 1967 he gave no speeches. Instead Nixon toured Europe, meeting with political leaders and with Pope Paul VI, far away from the line of fire.

By October of that year, though, Nixon was ready to wade into the waters of the Vietnam War again. In an article for the journal *Foreign Affairs* titled "Asia after Vietnam," Nixon sounded more skeptical about the war than ever. He warned that the continued conflict had given Americans a "distorted picture of Asia." Vietnam, "a small country on the rim of the continent, has filled the screen of our minds; but it does not fill the map." The implication? Vietnam wasn't worth a war.

With polls showing Nixon as the front-running Republican candidate for 1968, he ratcheted up his political profile, sending an "open letter" to the citizens of New Hampshire echoing the theme of national unity. Striking the pose of a healer, he told the people that "the Nation is in grave difficulties . . . The choices we face are larger than any differences among Republicans or among Democrats, larger even than the differences between the parties. . . . For these critical years, America needs new leadership. . . . During fourteen years in Washington . . . I believe I have found some answers." What they were, though, he wasn't saying.

Nixon fixed on the idea of "unity," invoking it as a theme that let him avoid taking positions on the issues. He used the word the way a Freudian psychiatrist uses his deliberately blank face, permitting people to read into it whatever interpretation his patient needs emotionally. To the left, "unity" meant ending the ostracism and persecution of antiwar students, and healing the divisions the war had caused. To the middle, "unity" meant ending the extremism of escalation on the one hand and draft resistance on the other. To the right, "unity" meant hunkering down, winning the war, and smashing the domestic opposition.

Determined to capture the middle, Nixon offered his blank-slate presence as a Republican third way—more moderate than the Goldwater-Reagan wing of the party, but less liberal than Nelson Rockefeller. Nixon speechwriter Ray Price said it best: "Nixon is neither a conservative nor a liberal. He is a Centrist." The hawk had flown to the center to nest.

Nixon's chief opponent for the Republican nomination was Governor George Romney of Michigan, the early front runner. A moderate Republican, he soon foundered when he remarked that he'd been "brainwashed" on an earlier inspection tour of Vietnam. Nixon's other rival, Nelson Rockefeller, proved too liberal for conservatives on domestic issues and too conservative for liberals on Vietnam. Without strong opposition, Nixon rolled to victory in the primaries, and wrapped up the GOP nomination well before the

Republican Convention. In his acceptance speech, Nixon returned to his call for national unity—without saying much about what principles he hoped to unite America behind. "We're going to win because at a time that America cries out for the unity that this administration has destroyed, the Republican party, after a spirited contest for its nomination for President and Vice President, stands united before the nation tonight. . . . A party that can unite itself will unite America."

By 1968, Americans were eager to find a candidate who would end the bloodletting in Vietnam and on America's streets and campuses. When Nixon applied for the job, they were so desperate for relief that they overlooked their natural reservations and flocked to his side. Without seriously scrutinizing Nixon's past, voters, disgusted by the president's failure to end the war in Vietnam, pledged to vote for anyone but Johnson.

Meanwhile, the Democratic Party was destroying itself. Johnson himself was a severely weakened president, yet the machinery remained firmly under his control. As a result, a vast number of antiwar Democrats felt bereft, without a candidate. Bobby Kennedy, whose constant criticism of the war made him a natural challenger to Johnson, sat on his hands, convinced that a challenge to the renomination of an incumbent president was doomed to failure.

Into the vacuum stepped, or rather ambled, the quixotic antiwar senator Eugene McCarthy of Minnesota. A diffident, deeply religious, philosophical candidate, he at first appeared to be no match for the political master in the White House. But as the war heated up with the North Vietnamese Tet Offensive in January 1968, McCarthy gathered steam. He ended up shocking the nation by winning more than 40 percent of the vote in the New Hampshire primary, an incredible performance for a virtual unknown against a sitting president.

Emboldened by McCarthy's showing, Kennedy finally entered the race on a strong antiwar platform. Johnson backed down fast when he saw Kennedy coming at him. Two weeks after Kennedy's announcement, Johnson told an amazed nation that "I shall not seek and I will not accept the nomination of my party for another term as your president."

After Johnson's withdrawal, his emasculated vice president,

Hubert H. Humphrey, entered the race. Once the beau ideal of the left in his days as a Senate civil rights advocate, Humphrey had ruined his liberal credibility by toeing his boss's line on Vietnam ever since Johnson made him vice president in 1964. Now Humphrey was positioning himself to stop Kennedy and win the Democratic nomination on behalf of the party's establishment.

But Humphrey's candidacy met with an embittered reaction from the young antiwar activists who had boosted McCarthy and Kennedy into contention. The struggle was hardly lost on Nixon, who watched the whole affair with rapt attention. As Nixon reported in his memoirs, "At one point [Humphrey] was driven almost to tears in front of the television cameras when he could not finish a speech before a derisive audience."

The Democratic Party was in freefall, weighted down by years of an unpopular war, and unexpectedly left without a clear leader. With Eugene McCarthy and Bobby Kennedy flanking him on the left, Hubert Humphrey was suddenly transformed in voters' eyes from a liberal ideologue to an improbable hawk.

Conscious that he would lose any direct contest to Kennedy, Humphrey refused to enter any of the primaries, counting on the delegations that were controlled by Democratic bosses to win the nomination. Since most states did not yet have laws requiring primaries, it was quite possible to avoid them and still prevail. It was *entering* them that could get you in trouble.

Lyndon Johnson, by now almost demented in his megalomaniacal determination to stay in Vietnam, held Humphrey's feet to the fire, demanding that he support the war as the price of his backing. Without Johnson, Humphrey wouldn't have the party bosses, and without the bosses he wouldn't win the nomination. So Humphrey submitted and continued to back the war.

In April 1968, Martin Luther King, Jr., was shot in Memphis, Tennessee. Black America exploded in grief, pain, and rage. Riots engulfed dozens of American cities, and once-peaceful neighborhoods became battlegrounds. Their bloody protests gave Nixon's calls for unity yet another meaning—in this context a subtle tinge of racism, the implication that white America should "unite" in the face of black protest.

Things grew even worse two months later, when Robert Kennedy was killed while he celebrated his victory in the pivotal California primary. As the nation's mourning period was extended yet again, Humphrey, who had not deigned to compete in any primary, parlayed the support of the Democratic political bosses into a first-ballot nomination at the convention in Chicago.

When it became clear that the Democratic Party was nominating Lyndon Johnson's hand-picked successor—despite the massive rejection the voters had given Mr. Johnson's war, the students who had animated the antiwar crusade escalated their battle as they demonstrated outside the Democratic convention hall in Chicago. The nation watched in horror as Chicago's police clubbed, kicked, beat, gassed, and arrested the largely peaceful—albeit noisy and impolite—demonstrators. The melee assured that the splits that had torn the Democrats asunder during the primaries would not quickly heal.

Meanwhile, Nixon continued to avoid taking a stand on the war, evading questions and mouthing generalities. Asked about Vietnam in May of 1968, he spoke vaguely about "American disillusionment" and expressed carefully couched reservations about the war. "We simply cannot continue . . . to carry this immense burden of helping small nations who come under attack, either externally or internally, without more assistance from other nations . . ." he remarked. "We need a new type of collective security arrangement in which the nations in the area would assume the primary responsibility of coming to the aid of a neighboring nation rather than calling upon the United States. . . ."

As Nixon refined his stance on Vietnam, Jonathan Aitken notes in *Nixon: A Life*, he "quietly dropped his earlier talk of escalation for military victory. In an important speech in Hampton, New Hampshire, he spoke of the search for 'honorable peace' (an important semantic change from his familiar phrase 'victorious peace')."

Nixon dropped more hints of his ambivalence toward the war in his August speech accepting the Republican nomination. "I am proud to have served in an administration which ended one war and kept the nation out of other wars for eight years afterward," he declared; his message was unmistakable.

Nixon let everyone know that he wanted to end the war, but remained purposefully elusive about how he planned to go about it. As early as March 1968 he was openly claiming to have a "plan to end the war" but refused to reveal it. It was a celebrated—and controversial—boast. The media claimed that Nixon had actually used the phrase "secret plan"; according to Herbert Parmet in *Richard Nixon and His America*, there is some evidence that the candidate himself never actually used the term, and that it came from a questioner at a campaign stop in New Hampshire. No matter: It came to be widely accepted that Nixon was keeping his plan under wraps.

Was he bluffing? In historian James Humes's estimation, "although Nixon's critics viewed the plan as a 'cynical campaign ploy,' Nixon did, indeed, have a plan—which he revealed only to intimates. It was to 'Vietnamize' the war by gradually withdrawing U.S. troops." In the end, he did just that—very, *very* slowly withdrawing them, while dragging out the war for four more years.

As Nixon later recalled, his plan "was not secret. Instead of reading my lips, you read my record . . . in the future, we should furnish the arms, we should furnish the aid, but they must furnish most of the men. That was Vietnamization."

But if the plan was not secret, it was certainly hard to find amid the chaff of Nixon's motherhood-and-apple-pie posturing about unity.

Why did Americans allow Nixon this ambiguity? Because we were a nation in agony. The assassinations of King and Kennedy, the death toll from the war, and the sense of national division—even revolution—in the air caused a national angst that was hard to bear. In five years, Americans had lost a president they loved to a bullet, and had come to feel an alienation and antipathy toward his successor the likes of which the nation had not experienced since the days of Herbert Hoover and the Great Depression. As Johnson lied his way into escalation after escalation, journalists invented a new term—"credibility gap"—to describe the nation's growing distrust of the chief executive.

Amid this anguish, Richard Nixon perceived that the only way to win the public's favor in such a delicate time was to hint at a solution to the nation's dilemma without ever saying out loud what the solution might be.

Whereas Lincoln divided the Democrats as with an explicit condemnation of slavery, Nixon divided them in 1968 using the suggestive power of implication. Lincoln's advocacy of a strong liberal position on the key national issue of his day pulled his Democratic opponent Stephen Douglas to the left, making him unacceptable to the southern wing of his party. Nixon's silence, on the other hand, permitted him to move toward the middle, offering Americans what appeared to be a moderate alternative to the proven hawkishness of Johnson and Humphrey.

Meanwhile, another division in Democratic ranks emerged on the right. Angered by Johnson's and Humphrey's liberalism on civil rights, and infuriated by court decisions requiring school busing in what would turn out to be a vain bid for school integration, racist southern whites and no less racist northern blue-collar workers rallied to the Independent presidential candidacy of Alabama Governor George Wallace. Wallace had famously tried to block the integration of the University of Alabama in June 1963. "The confrontation in the courtyard of the University . . . overnight [made] George Wallace . . . a national figure." Running now as an Independent in the general election, he siphoned even more votes from the beleaguered Democratic ticket.

While Nixon postured, coy and opaque about Vietnam, the core of the Democratic Party turned on hapless Hubert Humphrey in rage at the continued prosecution of a deadly—and dead-end—war.

Humphrey seemed to get into trouble each time he opened his mouth. In September 1968, the vice president's statements on Vietnam policy led to an immediate rebuke from Secretary of State Dean Rusk, likely at the behest of Johnson. In the words of journalist R. W. Apple, Humphrey was like "a man with his fingers stuck in taffy" as he tried to disentangle himself from the administration's Vietnam policy.

Desperate to bring his shattered party together, Humphrey called on every party leader he could—Ted Kennedy, George McGovern, Adlai Stevenson, Jr. But Humphrey was still attacked from the left over the war; Theodore H. White wrote afterward that he "felt [Humphrey's] cause hopeless," after a depressing interview with the candidate. According to White, the Democratic National Headquarters in Washington had become a "Crossroads of Lamentation."

To put distance between his positions and those of the administration of which he was a part, Humphrey delivered a nationally televised speech from Salt Lake City on September 30, 1968, and declared that his first priority as president would be to end the war. He said he felt that a halt in the bombing of North Vietnam—

demanded by the left—was "an acceptable risk for peace." The *New York Times* said the speech "may be a frail straw for doves to clutch at."

Nixon refused to debate Humphrey, which drove the Democrat crazy. Humphrey took to calling his rival "Richard the chicken-hearted." Emboldened by a basically positive reaction to his Vietnam speech, Humphrey kept issuing debate challenges, but Nixon stayed aloof. "I was determined," Nixon wrote, "not to be lured into a confrontation. It was not fear but self-interest that determined my decision on the debates. Naturally, my unwillingness to debate gave Humphrey a major campaign issue." Campaigning only in carefully controlled venues like tightly scripted televised "town meetings," Nixon took no unplanned questions and gave no answers he didn't want to give.

Why did reporters let Nixon get away with it? The media in the late 1960s were different and more passive than they are today. Reporters let candidates keep them at bay; the press had not yet learned that it could insist that candidates submit to the harsh questioning that would become normal later in the century.

As Nixon campaigned in the Midwest, he spotted a young girl holding a sign that read BRING US TOGETHER AGAIN. Ever the opportunist, the experienced politician pounced on the story and made it his campaign theme, claiming that his goal was to "bring us together." It was the ultimate expression of his call for unity, its effectiveness only enhanced by its lack of specificity.

George Wallace was particularly useful to Nixon, giving disaffected Democrats an alternative to the party of their fathers. Particularly in the South, where millions of people routinely and habitually voted Democratic, the Wallace candidacy would siphon votes from Humphrey.

In vain, Humphrey and those Democrats who stayed loyal tried to warn that Nixon's moderate posture was a ruse. Humphrey's vice presidential nominee, Senator Edmund Muskie of Maine, asked, "When did you start to get so progressive, Mr. Nixon? All your life you stood there and resisted and fought. You called my party the party of treason. . . . He fought Harry Truman. He fought Roosevelt. He fought Kennedy and Stevenson. He fought Lyndon Johnson. . . . Mr. Republican is saying he's a friend of the workingman. Now that's news for you, I'll guarantee you that—if he is a friend of the workingman, Scrooge is Santa Claus."

To reply to Humphrey's attack, Nixon decided to give ten nighttime radio broadcasts, assuming a serious, almost scholarly tone as he discussed such issues as welfare, youth, education, arms, peace. The speeches were a studied showcase for the "new Nixon." Gone was the divisiveness which had been Nixon's métier in previous campaigns. No thrust and slash this time. Unity was Nixon's watchword; he seemed to want nothing more than to apply balm to the nation's wounds. It was only in the light of the bitter paranoia that would dominate his presidency, the us against them psychology that led to his resignation from office, that the Nixon campaign of 1968 seems disingenuous.

Frustrated to distraction by the Democrats' inability to pin down the elusive Nixon, President Johnson decided to spring a last-minute surprise on the Republicans. On October 31, 1968, three days before the election, Johnson announced that "he had decided to call a total bombing halt over North Vietnam." Even Nixon was thrown: "I thought to myself that whatever this meant to North Vietnam, he had just dropped a pretty good bomb in the middle of my campaign." It was a clever move, to say the least. "Johnson was making the one move that I thought could determine the outcome of the election. Had I done all this work and come all this way only to be undermined by the powers of an incumbent who had decided against seeking reelection?"

But Nixon had a secret plan for this contingency, too. Henry A. Kissinger, the future secretary of state who served as Nixon's brilliant foreign policy advisor during the campaign, had warned that Johnson might make such an announcement about Vietnam. Now Nixon hastened to put into action the counterplay he had prepared.

For decades, ever since General Douglas MacArthur had champed at the bit to carry the Korean War into China, Asian anticommunists had always trusted Republicans more than Democrats. Nobody epitomized this relationship better than Anna Chennault, the Chinese widow of General Claire Chennault, who had worked to bolster the anticommunist regime in China.

In the spring of 1967, Nixon asked her to be his advisor on South-east Asian affairs. As Herbert Parmet writes, "cables intercepted by the National Security Agency from the South Vietnamese ambassador in Washington revealed that 'the Nixon entourage', working through Anna Chennault, was pressuring [South Vietnamese President] Thieu to resist the peace talks. Saigon, he was reminded, should hold out for a new administration, presumably one headed by Nixon . . ." As Parmet puts it, "Nixon was fighting back . . . he wanted to win and the dynamics of the campaign overwhelmed any other consideration."

Realizing what Nixon was up to, President Johnson ordered an FBI wiretap on Chennault's phone. Yet whatever evidence the tap may have turned up went unreleased, since it would put the Democrats in an "embarrassing position" to admit that they had been using wiretaps to defeat Nixon.

Heeding the signal from Nixon that he did not need to negotiate with the communists, South Vietnam's President Thieu cooled talk of a peace initiative and balked at negotiations. Nixon aides began to predict that Johnson's bombing halt would boomerang. Nixon told a Texas rally that "the prospects for peace are not as bright as they looked only a few days ago." But the prospects for his election were brighter.

Nixon had dodged a bullet. Humphrey briefly took the lead, but when South Vietnam would not agree to terms and the talks collapsed, Nixon moved ahead once again. As his final shot, Nixon charged that the bombing pause had allowed the North Vietnamese to move "thousands of supplies down the Ho Chi Minh trail, and our bombers are not able to stop them."

Nixon won. His victory was a close one—Nixon won 43.3 percent of the vote, to Humphrey's 42.6 percent and Wallace's 13.5 percent. But if Nixon had not managed to exploit the divisions within the opposing party, it would not have been a victory at all. In 1860, Abraham Lincoln had pulled his moderate opponent to the left, severing him from the South on the slavery issue. In 1968, Richard Nixon divided his enemies by pushing them apart, letting the left and right arms of the Democratic Party continue their fratricidal fight while Nixon waited quietly, in what appeared to be the middle. The more clearly Lincoln articulated his animosity toward slavery and his devotion to the Union, the more he dragged Douglas to the left to compete for northern votes, a position that rendered the Democrat unacceptable to the South. For Nixon, on the other hand, clarity was the antithesis of his battle strategy. Nixon was in fact determined to bomb and escalate his way to victory in Vietnam. But he concealed his true intentions in order to win desperate liberal Democrats over to his column.

Nixon's conduct in the 1968 election suggests one of the darker sides of politics—that manipulation and deceit can sometimes be effective in the game of divide and conquer.

DEWEY SPLINTERS THE DEMOCRATS BUT TRUMAN WINS ANYWAY

No American presidential candidate has ever faced the divisions within his own party that Harry S. Truman did in 1948—and lived to tell about it. Though he was the heir to the New Deal coalition of Franklin Delano Roosevelt, when Truman ran for president he found his party shattered into three pieces. To the left the liberals had deserted him, backing FDR's former vice president, Henry A. Wallace. To the right, the Southern racists abandoned him to back the young South Carolina governor Strom Thurmond. At the center, standing alone, was Harry Truman.

Yet somehow Truman managed to bounce back, using the division in his ranks to empower him. Shed of the Left and the Right, he preempted the center—and scored an upset two-million-vote victory in the 1948 elections.

How did he do it?

The divisions that beset Truman were the result of a deliberate effort by the Republicans to divide and conquer the Democrats. Determined to fracture the coalition that had vanquished them in the last four presidential elections, they conjured a strategy curiously similar to that which Abraham Lincoln had used in 1860, where Lincoln's strong position against slavery dragged Douglas so far to the left he lost the South. Republican candidate Thomas E. Dewey and his party tried to use their strong posture against domestic communism to force Truman so far to the right that he'd lose the backing of his party's left wing.

As the Democratic Left deserted Truman and rallied to the banner of Henry Wallace, Truman needed to attract liberals by embracing a strongly populist domestic policy, particularly on the issue of civil rights. But nothing went right for poor Harry Truman. His support for civil rights attracted African Americans and liberals, but so antagonized southern Democrats that they, too, bolted the party, joining Thurmond's Dixiecrats.

With the incumbent's constituency thus fractured, the presidency seemed ready for plucking by the Republican Party. Yet Truman confounded everyone by winning reelection. His ranks were divided as badly as those of Stephen Douglas or Hubert Humphrey, but he was not conquered. Indeed, Truman not only survived divisions; he seemed to thrive because of them.

How?

The short answer is that the divisions Truman faced empowered him by purifying his party. Without the southern racists, he was free to appeal overtly and successfully to northern blacks. Without the largely procommunist Left, he could attack Stalin and stand up to the Soviets in a way that robbed the Republicans of their best issue. The divisions Douglas and Humphrey faced depleted their ranks, and resolve. Truman's divisions only strengthened his.

As Roosevelt's successor, Harry S. Truman seemed like an inadequate replacement for a giant as he took the oath of office as president in 1945. Virtually unknown, Truman had served as vice president for little more than a month before Roosevelt died.

Short, physically unimpressive, and uneducated—he had never been to college—Truman's sharp and screechy voice contrasted with Roosevelt's rich baritone, and sounded terrible over the radio—a technology that had become FDR's lifeline to the country. The first years of Truman's presidency were a disaster.

The end of World War II was, in many ways, an economic boom time, but it also spelled chaos in the short term. When rigid wartime wage and price controls were lifted, the pent-up demand sent inflation soaring. With this discipline gone, Truman was confronted with "the longest, most costly siege of labor trouble in the nation's history."

Strikes crippled key industries: coal, steel, and cars. Meat grew scarce in 1946 as cattle ranchers staged a strike, demanding higher prices. Under pressure from America's housewives, Truman lifted meat price controls in October 1946, further aggravating inflation.

The coup de grâce was a railroad strike in May 1946, which paralyzed the nation. Grain and produce shipments rotted. Travelers were stranded. President Truman's own daughter, attending a play in New York, "had to borrow a car" to get back to Washington.

The beleaguered president asked Congress for permission to draft the striking railroad workers into the army. When the union finally caved into his pressure, labor leaders—and even Eleanor Roosevelt—denounced Truman as a strikebreaker. Vowing revenge, A. F. Whitney, president of the Brotherhood of Railroad Trainmen, summed up the national sense of Truman's inadequacy, sniping at the former haberdasher: "You can't make a President out of a ribbon clerk."

Abroad the news was even worse, as Soviet dictator Joseph Stalin gobbled up Poland, Hungary, Romania, Bulgaria, Czechoslovakia, Yugoslavia, and Albania, setting up governments controlled by Moscow. Americans became deeply concerned over the real threat of communism overseas—and the perceived danger of leftist infiltration at home. The Republican Party recognized that the specter of domestic communism, if treated correctly, could be a source of tremendous political capital for their side.

Franklin Roosevelt's Democratic Party, in fact, had more than its share of communist sympathizers. Their attraction to communism was not surprising. The Great Depression of the 1930s had severely tarnished the image of capitalism. The economy of Soviet Russia seemed to be surging ahead while the West stagnated. Word had not yet leaked out about the death factory Stalin was running in Russia, which had consumed twenty million lives. The Soviet Union's heroic defense against Hitler's armies only added to its ranks of admirers in the United States.

But significant numbers of Americans were growing increasingly worried about communist expansion abroad. Republicans soon began fanning fears of leftist infiltration at home, charging that communist sympathizers held top positions in the U.S. government— and that Truman was doing nothing to throw them out. Just months into the Truman administration, in November 1945, the U.S. ambassador to China, Patrick J. Hurley, resigned and announced that the State Department was favoring the Chinese communists.

As Truman biographer David McCullough writes, "a red scare was clearly on the rise." Republicans were accusing Truman of "pursuing a policy of appeasing the Russians abroad and fostering communism at home." Republican Senator Robert Taft said the Democratic Party was "so divided between communism and Americanism that its foreign policy can only be futile and contradictory and make the United States the laughing stock of the world."

Truman's ratings plummeted, dropping to 32 percent as the Democratic Party suffered its worst defeat in twenty years in the congressional elections of 1946. Republicans captured both houses of Congress and a majority of state governorships. Likely Republican presidential nominee Thomas E. Dewey was reelected governor of New York by the largest margin ever.

Delighted with their success in exploiting the Red Scare, the Republicans continued their offensive. The House of Representatives' Committee on Un-American Activities, now animated by the work of freshman Congressman Richard M. Nixon, pounded away, stoking fears of domestic and international communism. As biographer Richard Norton Smith notes, Dewey "implored his countrymen," as Stalin grabbed Eastern Europe, "to behave like 'hardheaded Americans instead of softheaded saps' in confronting Russian aggression in Europe."

Dewey singled out Truman for scorn, charging him with inaction in the face of Soviet expansion. Communism was advancing steadily, he charged; "the tragic fact is that too often our own government seems so far to have lost faith in our system of free opportunity as to encourage this Communist advance, not hinder it." Warning of the danger of domestic communism, he said that "Communists and fellow travelers [have] risen to positions of trust in our government."

And Dewey was a moderate. The leader of the Republican right wing, Senator Robert Taft, embodied the faction's isolationism. *Fortune* magazine accurately depicted his as "one of that vast group of Americans to whom other countries seem merely odd places, full of uncertain plumbing, funny-colored money, and people talking languages one can't understand."

When Truman "accepted a reporter's characterization of the House [Un-American Activities Committee's] work as a 'red herring,' " the Republicans attacked him all the more. The domestic communism issue soon began taking its toll on Truman's popularity; the 1946 congressional defeat was blamed by many on the GOP's campaign to tar the Democrats as soft on communism.

To deflect Republican criticism, Truman moved to the right on the issue, establishing a Federal Employees Loyalty and Security Program in March 1947 that would allow for the dismissal of federal employees based on investigations into their "loyalty." Justifying the program, Truman tried to separate his initiative from the Republican charges that had inspired it: "I am not worried about the Communist Party taking over the government," he said, "but I am against a person, whose loyalty is not to the government of the United States, holding a government job."

Four years later, by 1951, three million federal employees had been investigated, and thousands were forced to resign—although only 212 were actually fired.

Aide Clark Clifford later said that Truman thought the fears of communist subversion were "baloney," but that "political pressures were such that he had to recognize it." Truman's action had the desired effect of appeasing the right wing, as *Time* correspondent Frank McNaughton reported, "Republicans are now taking Truman seriously . . . [his] order to root out subversives from government employment hit a solid note with Congress, and further pulled the rug from under his political detractors." To McNaughton it seemed the stroke had worked: "The Republicans are beginning to realize Truman is no pushover."

But Truman's move alienated his party's liberals—as the Republicans doubtless hoped it would. The Left greeted Truman's rightward drift with suspicion. Fanned by resentment of Truman's tough action against labor strikes, it began to wonder if Truman was worthy of its support.

Meanwhile, Truman was both alert and aggressive in dealing with the real source of the communist threat—Stalin himself. After a brief carryover of wartime warmth for the Russian leader, Truman—like most Americans—became seriously disillusioned with Stalin. Concerned that the Soviets wanted to add Iran, Greece, and Turkey to their empire, Truman decided to get tough with Stalin. "I'm tired of babying the Soviets," Truman fumed. "They only understand one language: 'how many divisions have you?'"

Truman biographer David McCullough notes that almost overnight "relations with Russia were deteriorating." Truman knew that he had to draw a line to stop Stalin's aggression, and the opportunity came in Greece, where a Soviet-inspired civil war raged. As Truman recalled, "Under Soviet direction . . . Greece's northern neighbors—Yugoslavia, Bulgaria, and Albania—were conducting a drive to establish a Communist Greece."

Truman decided to send massive military and economic aid to both Greece and Turkey to forestall communist advances. Addressing Congress on March 12, 1947, he warned that "the seeds of totalitarian regimes . . . are nurtured by misery and want . . . the free peoples of the world look to us for support in maintaining their freedoms."

As soon as they were checked in Greece, the Soviets ratcheted up the pressure on the Allies in Berlin. The former German capital, divided into a democratic West and a communist East, was located deep in East Germany. Its western half had to be supplied by Allied convoys originating in West Germany that ran through the communist territory into West Berlin.

Now, in an effort to starve West Berlin into submission, Stalin ordered the supply convoys blocked. Daringly, Truman decided to resupply West Berlin from the air. Planes took off every few minutes, bringing food, fuel, and other vital supplies to the people of West Berlin. Incredibly, the airlift worked, and once again the Soviets were thwarted.

But Truman's tough line against Stalin—and his domestic loyalty-security program—were too much for the Democratic Party's left wing to stomach. Soon, from its ranks, emerged a leader to challenge Truman: Henry A. Wallace.

Wallace had been part of the New Deal from its inception. In the 1930s he had served as Roosevelt's Secretary of Agriculture, helping to mastermind the New Deal farm programs that rescued American farmers from bankruptcy by limiting production and supporting prices. A dedicated liberal, he was one of the best-liked members of the New Deal cabinet.

When James Garner, FDR's vice president since 1933, objected to Roosevelt's policies, FDR dumped him from the ticket in 1940 and replaced him with Wallace, who was, in turn, himself dumped to make way for Truman in 1944. Indeed, had moderate Democrats not prevailed on Roosevelt to demote Wallace back to the cabinet, Wallace and not Truman would have assumed the presidency upon FDR's death in 1945—no doubt spelling a very different postwar world order.

Wallace's resentment at being axed from the ticket burned brightly, and he came increasingly under the sway of communist influences. As Truman confronted Stalin in Europe, Wallace grew more and more restive.

Soon he began to stray from the reservation. In 1946, while still serving as Truman's secretary of commerce, Wallace addressed a political rally at Madison Square Garden in New York, criticizing Truman's strong line with the Soviets. "The tougher we get with Russia," Wallace said, "the tougher they will get with us." Barraged with criticism from within the administration, Wallace resigned from the cabinet.

His differences with Truman grew wider. When the president asked Congress to stop communism in Greece and Turkey, Wallace denounced the speech in a radio address sponsored by a group called the Progressive Citizens of America. "The world is hungry and insecure," Wallace said, "and the people of all lands demand change. American loans for military purposes won't stop them. . . . America will become the most hated nation in the world."

In April 1947 Wallace toured Europe, denouncing Truman's plan of aid for Greece and Turkey. Back in the United States, he called Truman's policy—which had come to be known as the Truman Doctrine—"a curious mixture of power politics and international carpetbagging," and said, "I am not afraid of communism."

As the cold war grew more confrontational in 1947, Wallace hinted at defection from Truman's ranks in a Los Angeles speech. "If the Democratic Party departs from the ideals of Franklin D. Roosevelt," he pledged, "I shall desert altogether from that party." Then he got his chance: meeting in New York on October 18, 1947, a group of labor leaders—along with Eugene Dennis, the general secretary of the United States Communist Party—decided to run Henry Wallace as a third-party candidate for president.

Under attack from the Republican Right and the Wallace Left, Truman's advisors pondered what course to take. The more conservative among them, led by Treasury Secretary John W. Snyder, counseled moving to the right to counter the Republican threat. But liberals, led by Clark Clifford, wanted him to tack left to win back the liberals who might otherwise defect to Wallace.

Truman vacillated. "At the center of an incessant tug of war," Irwin Ross writes, "he let himself be pulled first in one direction, then in the other, depending on the varying pressures, at any moment, of expediency [and] friendship." As Clifford recalled, "most of the Cabinet and congressional leaders were urging Mr. Truman to go slow, to veer a little closer to the conservative line. They held the image of Bob Taft before him like a bogeyman. We were pushing him the other way."

The president began to tilt toward the left. The move followed his own liberal tendencies, and Clifford, who was from Truman's home state of Missouri, was an encouraging ally. Clifford gained in power, and by mid-1947 was Truman's most trusted advisor. Clifford urged Truman to adopt a program of liberal economic measures and civil rights reforms, and called on Truman to appoint more progressives to high levels in the government.

The turning point came on November 19, 1947, when Clifford wrote a forty-three-page memo to Truman calling for a decisive move to the left. Anticipating that "Henry Wallace will be the candidate of a third party," he explained that "from the Communist long range point of view there is nothing to lose and much to gain if a Republican becomes the next President. The best way it can help achieve that result . . . is to split the Independent and labor union vote between President Truman and Wallace—and thus ensure the Republican candidate's election."

With Wallace in the race, Clifford assumed that "the independent and progressive voter will hold the balance of power in 1948." But, he warned, liberals "will not actively support President Truman unless a great effort is made."

Clifford's key proposal was that Truman take a strong position on civil rights. Because of the concentration of black voters in key northern states, Clifford wrote, "The Negro vote *does* hold the balance of power."

Since the 1860s, southern Democratic racism and the memory of Abraham Lincoln had led blacks to vote Republican, but Roosevelt's vigorous courtship of African American voters—through largely symbolic gestures—had won many over to the Democratic side.

Clifford warned that Dewey and the Republicans could win back black voters because of his "assiduous and continuous cultivation of the New York Negro vote . . . and his insistence that his legislature pass a state anti-discrimination act." He noted that New York Democratic political boss Ed Flynn had privately predicted Dewey would carry New York—the largest block of electoral votes in the country— "because he controls the Negro and Italian blocs."

Clifford warned that "the northern Negro is today ready to swing back to his traditional moorings—the Republican Party," and that "unless the administration makes a determined campaign to help the Negro . . . the Negro vote is already lost."

But Clifford made a crucial mistake in his calculation that would have profound consequences. He dismissed the notion that a liberal civil rights stand might induce yet another fissure in Democratic ranks by driving southerners out of the party. "It is inconceivable," Clifford wrote, "that any policies initiated by the Truman Administration no matter how 'liberal' could so alienate the south in the next year that it would revolt. As always, the south can be considered safely Democratic." Famous last words!

Following Clifford's advice, Truman began to push civil rights. On February 2, 1948, in a landmark address to Congress, he urged "modern, comprehensive civil rights laws, adequate to the needs of the day." Irwin Ross wrote that he demanded "anti-lynching legislation, an end to segregation in interstate transportation, the establishment of a Fair Employment Practices Commission . . . stronger statutory protection for the right to vote . . . and the establishment of a . . . Commission on Civil Rights."

The South reacted quickly and furiously. South Carolina governor Strom Thurmond (at this writing still in the Senate) called Truman's civil rights message "the most astounding presidential message in political history," and credited it with the birth of the so-called "states' rights movement."

Truman wouldn't budge in the face of Thurmond's attack. "The leaders of 'white supremacy,'" he later wrote, "began at once their campaign of demagoguery." The southern rebellion spread quickly: At the Conference of Southern Governors in Florida, a few days after Truman's speech, Governor Fielding L. Wright of Mississippi suggested forming a Southern political conference to "bolt" from the national party. At a Democratic dinner in Little Rock, Arkansas, hundreds of attendees rose and left the room when Truman's voice came on the loudspeaker. "Four days later, on February 23, [1948,] a committee of the Conference of Southern Governors . . . led by . . . Thurmond . . . [met to consider] the southerners' contention that the President's proposals were an unconstitutional invasion of states' rights."

When a reporter pointed out to Thurmond that "Truman is only following the platform that Roosevelt advocated," Thurmond replied, "I agree but Truman really means it."

Thurmond asked Democratic Party Chairman McGrath to use his "influence . . . to have the highly controversial civil rights legislation, which tends to divide our people, withdrawn from consideration by the Congress." McGrath refused. Enraged, the southern governors sent out the word that "the southern states are aroused," and warned Truman not to assume the once-solid South was "in the bag" for the Democrats.

When Truman insisted that the Democratic Party platform in 1948 endorse strong civil rights legislation, a split became inevitable. As Truman recalled, "I did not discount the handicap which the loss of a 'Solid South' presented. . . . I knew that it might mean the difference between victory and defeat in November. . . . I knew, too, that if I deserted the civil liberties plank of the Democratic party platform I could heal the breach, but I have never traded principles for votes and I did not intend to start . . . regardless of how it might affect the election."

As Dewey's campaign manager Herb Brownell remembered, "a four party race was in prospect—really a Republican against a splintered Democratic party."

Divisions in the camp of your rival don't just happen by accident. They are a product of your own strategy and your own positioning. Every action has an equal and opposite reaction. When the Republicans attacked domestic communism, they forced Truman to the right on the issue, which sent Henry Wallace and the liberals packing.

Then, to keep the Left in line, Truman pushed civil rights. That, in turn, drove the southerners out. The Democratic Party was like a projectile that had shattered upon impact with the Republican challenge.

It was the same kind of move Lincoln had made against Douglas, attacking him over slavery to force him to the left and make him lose the South.

But the lesson of the Truman campaign is that divisions can help your cause.

To most observers, Truman seemed dead. But he realized what his opponents missed: that the divisions he faced empowered him, by purifying his supporters. Without the base alloys of communist or racist support, Truman was free to speak clearly for freedom and justice at home and abroad.

As Truman later recalled, "The greatest achievement was winning without the extreme radicals in the party and without the Solid South. It is customary for a politician to say that he wants all the votes he can get, but I was happy . . . to be elected . . . by a Democratic party that did not depend upon either the extreme left-wing or the southern bloc."

With Clifford's assistance, Truman hit on a novel strategy to heal the breaches in his party. Instead of trying to win the Wallace-ites back by soft pedaling his opposition to communism or seeking to conciliate the South by backtracking on civil rights, he mixed and matched. He used civil rights to attract the liberals, and anticommunism to lure the conservatives.

It was an ingenious move, making a winning virtue of necessity. Taking hold of the new issues fate had dealt him, Truman began to line up important new constituencies to replace the ones he had lost. Where Douglas had tried to placate the South over slavery and Humphrey tried to salvage his relationship with liberals over Vietnam, Truman knew it was impossible to cave in to either Wallace or Thurmond.

To win back the Left, Truman took the offensive with the liberal domestic policy urged by the Clifford memorandum. In June 1947, after the Republican Congress passed the Taft-Hartley Act, which gave the president the power to order strikers to return to work for a "cooling off" period, Truman vetoed it. While it was enacted over his objection, his strong stand rallied labor to his cause. In the end, the unions' support was crucial in overcoming the liberal disaffection that had festered since he instituted his anticommunist loyalty programs.

Truman also lavished attention on farmers. Another key segment of the Democratic coalition, the agricultural bloc had staunchly supported Roosevelt as he battled the Depression, but began to defect to the Republicans in 1937. "By November 1946, [their defection] was complete," one retrospective history wrote; "farmers in the Midwest were expected to support overwhelmingly the Republican ticket in 1948."

But Truman knew how to get the farmers, by hitting on their old bugaboo—Wall Street. "The Wall Street reactionaries," Truman said, "are not satisfied with being rich . . . they want to increase their power . . . they are gluttons of privilege . . . the Republican strategy is to divide the farmer and the industrial worker—to get them squabbling with each other—so that big business can grasp the balance of power and take the country over, lock, stock, and barrel. . . . That's plain hokum. It's an old political trick. 'If you can't convince 'em, confuse 'em.' But this time it won't work."

With straight talk like that, Truman locked up the farm vote. As Dewey said, "You can analyze figures from now 'til kingdom come and all they will show is that we lost the farm vote which we had in 1944 and that lost the election."

But the farm vote would have meant little if Truman lost New York. As Clifford noted in his strategy memo, "except for Wilson in 1916, no candidate since 1876 has lost New York and won the presidency." And with New York's popular governor as the GOP nominee, the state was up for grabs. When it became clear that Wallace's progressives would desert Truman, the chances for carrying this pivotal—and liberal—state deteriorated further.

As Clifford had pointed out, the black vote was key to winning New York. Again, the very defections that threatened to sink Truman ended up empowering him. Freed of the need to appease the southern racists now that Thurmond had left the party, he was able to support civil rights vocally and vigorously. With virtually the entire Democratic Party organizations from South Carolina, Mississippi, Alabama, and Georgia supporting Thurmond, Truman no longer had to worry about conciliating the right wing of his party on race.

With no hope of carrying that bloc, he was empowered to go after the black vote in the north.

The other key to winning New York was its large Jewish vote. As the United Nations was preparing to establish the nation of Israel, Clifford urged Truman to move decisively to recognize the new state as soon as the United Nations voted to bring it into existence.

Former, current, and future secretaries of state and defense George C. Marshall, Dean Acheson, Robert Lovett, and Dean Rusk— a group of foreign policy leaders who subsequently became known as the "Wise Men"—were all dead set against it. As Clifford recalled in his memoirs, the State Department leaders did "everything in their power to prevent, thwart, or delay the president's Palestine policy in 1947 and 1948."

Following Clifford's wise counsel, Truman was the first to recognize Israel—on May 14, 1948, just in time for campaign season. The Soviet Union followed three days later. As Dewey's manager Herb Brownell said, "the move to recognize Israel firmed up Truman's support among liberal voters—some of whom had been leaning toward Wallace."

Wallace's candidacy also freed Truman of the need to appease the left in his own party by going easy on the Soviet Union. Now the president was poised to pitch a battle with Stalin, and reap political reward at home. Clifford pointed out that "Republican propaganda is repetitious on the theme that Soviet expansion in Europe could and should have been stopped long ago and that . . . only President Truman's actions at [the] Potsdam [Conference] prevented this from happening."

A strong policy against the Soviets, Clifford reasoned, would help Truman counter the Republican attacks and make the issue of communism work in his favor. "The nation is already united behind the President on this issue. The worse matters get . . . the more there is a sense of crisis. In times of crisis the American citizen tends to back up his President." Truman's anticommunist positioning robbed the Republicans of the key issue they had used to capture Congress in 1946.

The very splits in his party that seemed certain to sink Truman in 1948 were now magically realigned in his favor. Free to appeal aggressively to blacks, Jews, and farmers and to confront the Soviets abroad,

Truman was able to redraw the political map in his favor, cobbling together a new coalition that would carry him into office.

What's the lesson here? Among other things, that a divided party can recombine in new and more fruitful ways. Political opposition can fragment any given coalition—but the right kind of creative leadership can recapture their loyalty by rewriting the terms of the debate.

Politicians who kowtow to established political factions can often become hobbled by the association, as Truman clearly was before 1948. But Truman demonstrated that sometimes the most threatening kinds of political splits can actually be a boon—especially when they end up purging your camp of troublesome strains of extremism. With communists and racists to worry about, Truman would never have been free to take the kinds of positions he did.

Harry S. Truman, in other words, didn't let his friends hold him back. And though he may have lost a few along the way, the redoubled loyalty of those he retained was more than enough to prevail.

Truman not only won, but he had fun doing it. He was surprised when the Republicans adopted a liberal platform that called for increasing old-age pensions, equal pay for women, a ban on poll taxes (which were used to stop blacks from voting), an employment antidiscrimination law, low-cost housing, and support for small family farms. In light of Republican congressional opposition to these same programs, Truman saw the platform as an act of hypocrisy.

So Truman called their bluff. After the Republican Convention, he summoned the Congress back into special session and dared it to enact the provisions of its own party platform into law. July 26, the day the session began, was Turnip Day in Missouri; he joked that he was calling the Congress into session to see what turned up.

As Brownell remembers, "Clark Clifford . . . devised a clever campaign strategy for Truman. Clifford knew there were great differences between Dewey's moderate, internationalist position, and the much more conservative and isolationist positions of the Republican Congressional leadership. Therefore, Truman opened his campaign by calling the Republican-controlled Congress back into session and challenging them to pass legislation which supported the moderate Republican platform. The Republicans in Congress, of course, wouldn't do it."

"In retrospect," Brownell said, "the failure of the Republican leaders in Congress to accommodate their presidential candidate was damaging. The conservative Republicans in Congress had for too long been accustomed to attacking Roosevelt . . . taking essentially a negative stance." When it became clear that a more proactive approach was called for, the Congressional Republicans failed to meet the challenge.

Suddenly, Truman had switched opponents. "He sought to turn the campaign into a Truman-versus-Republican Congress contest rather than a Truman-Dewey race," Brownell wrote. "The rift in the Republican ranks [between Dewey's liberalism and the Congressional Republicans' conservatism] enabled Truman to depict the [Republican] Party in a negative light, resuscitating public suspicions of the Republicans as the party of Herbert Hoover and political reaction," Dewey's handler admitted ruefully. "The failures of the Republican-controlled Eightieth Congress—rather than Truman's own record—became the focus of the rest of the campaign, enabling Truman to take the offensive."

With the special session of Congress, Brownell reported, "basically, Truman turned populist in his appeals, pitting the poor against the rich, the farmers against Wall Street." Freed of the need to court his own party's conservatives by the Thurmond defection, Truman could aggressively go after the liberals. He had found his stride.

It's another lesson of Truman's campaign: Choose your own opponent. There's no need to spend all your time attacking the one who is running against you, the rival who's blocking your promotion. When Truman realized he'd have a hard time beating the liberal Dewey, he decided to run against the Republican Congress instead. It was a tactic Bill Clinton would echo in 1996, reserving his harshest barbs for House Speaker Newt Gingrich even though Bob Dole was his nominal opponent. Gingrich may not have been running, but he was such a tempting target that Clinton made him his adversary—and made Dole seem irrelevant in the process.

While Truman was positioning himself properly, Dewey was also proving to be one of the world's worst presidential candidates.

His prospects had seemed bright as he handily defeated Minnesota's Harold Stassen to win the Republican presidential nomination for the second time. "We thought Dewey would be elected,"

Brownell recalled, "and we based our predictions not so much on the Republicans' strengths as on the Democrats' weaknesses. Truman had some serious problems with which to contend." On the surface the secession of Wallace and Thurmond seemed fatal to Truman, and Dewey's camp was ecstatic. "There were really three Democratic candidates, and we believed this split in the Democratic party would ensure Dewey's victory."

But Brownell underestimated his own candidate's capacity to blow the election. Where Truman was energized and catalyzed by the splits in his own party, Dewey was emasculated by the splits between the right-wing congressional leaders and his own moderate-liberal wing. The split not only "prevented the party from moving in one direction on foreign policy in the 1940s, but forestalled the development of a . . . Republican alternative to the New Deal and the Fair Deal in domestic affairs."

But Republican infighting wasn't the Republicans' only problem. Another was Dewey himself. Clifford described the Republican's speaking style as "soporific," his speeches as "bland." And Dewey's own campaign advisor's verdict was much harsher: With his mustache, Brownell recalled, Dewey "looked like a devil, and many people judged him as a sinister figure." Paranoid about germs, Dewey hated shaking hands—surely in itself a mortal pathology for a presidential candidate. "He was candid with everyone," to the point of being abrasive and rude. He was pompous and aloof. "His wife didn't like politics and wanted him to get out."

When Brownell sent Dewey to a western rodeo to counter his image as an Eastern Wall Street dandy, he showed up wearing a homburg—a banker's formal hat, and universal symbol of the establishment. "Instead of the big cheer that we had expected," Brownell remembered, "he was booed, and the crowd completely overlooked the splendid speech he made on farm issues."

While Dewey did his best to avoid contact with the human race, Truman toured the country giving impromptu, extemporaneous, off-the-cuff speeches slashing and burning his GOP opposition. His campaign train stopped at stations so tiny that you had to blow the whistle on board to notify the conductor that you wanted to get off— thus the famous "whistle stop" tour. Delivering seventy-one speeches in thirty-three days, he spoke in a plain language people could understand, showing his wit and personality, and bringing them a message that they felt made sense.

Dewey, on the other hand, according to historian George H. Mayer, "adopted the classic strategy of saying nothing that would drive the two splinter parties back to the Democratic fold. He restricted himself to platitudes and innocuous appeals for national unity . . . [and] tried to take refuge in silence." Dewey was so sure he would beat Truman that on August 9, 1948, he was already musing to his mother, "I do not know about accommodations at the White House for the family."

Dewey became an object of ridicule when Alice Roosevelt Longworth, the tart-tongued daughter of revered president Theodore Roosevelt, was widely quoted referring to Dewey as "the bridegroom on top of the wedding cake." Her remark beautifully captured the prissy, primped, fastidious Wall Street Republican, and seemed to give expression to a latent reservation held by voters across America.

Truman, meanwhile, was becoming feistier with each passing day. He lashed out at Wallace and Thurmond, calling Democratic southern defectors "crackpots" and dismissing Wallace's supporters as "part of the contemptible communist minority." Before long the Wallace "Progressives" self-destructed, as the candidate's own communist sympathies eroded his support. As Mayer observes, in the end, his constituency included only "the Communist-infiltrated labor unions in the urban East and assorted intellectuals who favored a softer foreign policy toward the Soviet Union." United Automobile Workers president Walter Reuther described how the communists took over the Wallace campaign: "Communists perform the most complete valet service in the world. They write your speeches, they do your thinking for you . . . that's the trouble with Henry Wallace."

The final straw came when Wallace compared the communists to the early prophets of Christianity, volunteering that "the communists are the closest things to early Christian martyrs"—a comparison that might have surprised the lions. He further weakened his position by calling for a U.S. withdrawal from Berlin and by accepting the formal endorsement of the Communist Party. Against the backdrop of Stalin's increasingly blatant aggression in Europe, "Wallace's argument that American 'imperialism' was responsible for the Cold War" rang hollow.

As Wallace moved leftward, Truman's strategy of appealing to liberals was undercutting the Progressive candidate's political appeal. The president's "civil rights program, and his baiting of the Republican Congress effectively countered Wallace's thesis that there was no essential difference between the two major parties."

More and more liberals lined up against Wallace as the campaign unfolded. "The International Ladies Garment Workers Union denounced Wallace's candidacy as 'communist-inspired.'" More patronizing was Reuther's characterization of Wallace as a "lost soul." *The Nation*, a key liberal magazine, said that Wallace's "quixotic politics could help Dewey by splitting the Democrats."

With polls showing that 51 percent of voters were convinced that the Wallace party was run by communists, pollster George Gallup confirmed that this belief was "one of the reasons why the Wallace third party has not been able to increase its following."

As Wallace's candidacy became unglued, Thurmond's appeal became increasingly isolated to the deep South. His party's nickname—the Dixiecrats—signaled Thurmond's geographically limited appeal, and soon the threat seemed minor even in comparison with Wallace's foundering efforts.

In the end, the Wallace and Thurmond candidacies seemed actually to help Truman. It was Henry Wallace, not Harry Truman, who drew the wrath of anticommunist attacks—and Thurmond made Truman seem more attractive to black voters in northern cities. Wallace helped prove to Republican and independent voters that the president was no communist, and Thurmond showed blacks he was no racist. African American voters reasoned "anyone Thurmond attacked couldn't be all bad."

Despite Truman's obvious gains, the preelection polls still put Dewey ahead. Gallup had Dewey on top 49.5 percent to Truman's 44.5 percent. The Crossley Poll showed Dewey ahead by 49.9 to Truman's 44.8 percent. The Roper Organization stopped polling before election day since Dewey's victory seemed such a foregone conclusion.

Truman's victory, when it came, shocked political pundits everywhere. The Chicago *Tribune*'s famous, erroneous headline, "Dewey

Defeats Truman," was held aloft by a jubilant Truman—a mistake akin to the announcement by U.S. television networks that Al Gore had carried Florida and with it the 2000 presidential election.

Truman defeated Dewey in electoral votes by 303–189. In the popular vote, Truman got 49.5 percent to Dewey's 45 percent. Thurmond and Wallace each got roughly 2 percent of the vote; Wallace failed to carry a single state, and Thurmond carried only four—Alabama, Louisiana, Mississippi, and South Carolina, amounting to a paltry 38 electoral votes.

Democratic division crippled Stephen A. Douglas in 1860 and Hubert H. Humphrey in 1968. But in 1948 it empowered Harry Truman. Rather than waste his time trying to win back the defectors on their issues, he won them back on his own terms. Truman's Democratic Party was divided, but in the end it was he who conquered.

STRATEGY FOUR

REFORM YOUR OWN PARTY

When your party can't stop losing, there's really only one way to start winning again—you've got to change. It can be difficult to find a way to persuade the powers that be to surrender control and alter their ways, but nothing does that better than losing election after election after election.

The Democratic Party changed after losing repeatedly in 1980, 1984, and 1988 by nominating Bill Clinton and moving to the center. The Republicans changed after losing in 1992 and 1996 by backing George W. Bush and embracing "compassionate conservatism." A long time in the wilderness will do that to a political party and the interest groups that support it.

But what is truly extraordinary about the process of party reform is how it empowers the reformer, and can make his election almost inevitable. Slaying the dragons of one's own party can be such an attractive spectacle that independent voters looking in at the fratricide often end up flocking to the reformer in droves.

In America this is especially true, because while our politics is partisan, our voters are not. Everything in Washington, D.C., is either Democratic or Republican; all but two of the 535 members of Congress fit into these neat categories. It's surprising that the Capitol restrooms themselves aren't separated by party as well as by gender.

But while our nation's capital is sharply divided into two camps, voters in our time increasingly reject affiliation with either political organization. While 30 to 35 percent usually call themselves Democrats and 25 to 30 percent normally say they are Republicans, a plurality—40 percent—identify themselves as independents: a plague on both your houses. Each year they vote for some Democrats and for some Republicans, but they resist signing on formally with either party. They date, but they don't marry.

Why? Often it's because, while individual candidates may attract them, something about each party turns them off and prevents them from identifying with it. The Republican Party's conservative views on abortion and gun control are the decisive elements in limiting the party's appeal among Independent voters; the Democratic Party's longtime sponsorship by labor unions and minority groups, and its willingness to embrace their agenda to the exclusion of others, has a similar effect.

In the 2000 presidential campaign, the intransigence of the party elites to change was glaringly evident in the primaries, when Independents overwhelmingly supported Senator John McCain in the Republican primary and former Senator Bill Bradley in the Democratic primary. The party faithful, on the other hand, turned out heavily for Bush against McCain and for Gore against Bradley. But rather than reach out to these Independents, both Bush and Gore chose instead to hunker down and concentrate on winning the primaries by solidifying their support among the party faithful, with Bush warning them that McCain was unreliable in his opposition to abortion and Gore claiming that Bradley's health-care spending proposals were not sufficiently generous. The result, of course, was one of the least inspiring campaigns in recent memory.

Why is it so difficult for stalwart partisans to get the message and change? Because they value purity more than victory. Determined, at all costs, to preserve the elements of their platform, they would "rather be right [or left] than president." (It's worth noting that Henry Clay, the creator of that famous phrase, lost three presidential races.)

The dogmatism of the core of a political party can leave an ambitious new candidate with a basic dilemma: How does one appeal to enough Independents to prevail in the general election, while still preserving enough of his own base to win the party's nomination?

Trying to accommodate the desires of a trisected electorate—Democrat, Republican, and Independent—in a two-party system defines the central problem facing American politicians in their pursuit of power.

But some political figures throughout our history have learned to listen and listened to learn. These men and women have heard the grievances of those who criticized them and their party, and paid attention to voters who demanded changes. While many politicians have reflexively defended their party against these critiques, others have embraced them—and treated them not as obstacles but as catalysts to self-criticism and self-reform.

Undoubtedly, some did so opportunistically, eager to give the political customer what he wants in an effort to make a sale. Demagogues who stood for nothing but their own success—who "cut [their] conscience to fit this year's fashions," in the words of Lillian Hellman—are an inevitable fixture within a loosely organized political framework like ours.

But other leaders find that they have experienced a conversion more real, internalized, and profound. In their hearts, these political leaders shared something of the disappointment independent voters felt. Having absorbed the same lessons, they weren't shy about concluding that their own party was in need of basic reform.

Amid the partisan debate that swirls around American politics, with its fierce winds and extreme tides, such maverick figures paused to hear what their critics were saying. Instead of refuting the charges and attacks, they learned from them—and resolved to fix what was broken in their own political party.

This section focuses on three such men: Tony Blair in Britain, Junichiro Koizumi in Japan, and George McGovern in the United States. All three reformed their parties. Each slew its old guard. But only the first two won: McGovern succeeded in reforming his party, but he so alienated those he emasculated that they killed him in their dying act.

The trick is twofold: You've got to defeat the bad, old leader of your own party as you assume the mantle of leadership—and then make sure they don't knife you in the back as you try to carry your reformed party to victory.

What's the key to accomplishing this? For those who have been successful, it often means moving to the center in one's reform efforts. When you move to the center, where else are the extreme Left or the extreme Right of your party going to go? To the other side of the ideological spectrum? Not very likely.

When Tony Blair moved Britain's Labour Party to the center, the union bosses had no option but to follow him; after all, they could hardly embrace the Tories, their archenemies. When Clinton moved the Democrats to the center, the liberals followed duly along, kicking and screaming though they were. What else could they do? Vote with the Republican Right?

But this kind of maneuver can be trickier if the ideological lines between your parties aren't as clear. When George McGovern tried to move the Democratic Party to the left, it was an open invitation for the more conservative political bosses he opposed to defect to Nixon. As McGovern moved the Democrats to the left, he made the more moderate Republican platform far more attractive to his own party's conservatives. Rather than follow the treacherous McGovern, it suddenly seemed much more attractive to join forces with Nixon and defeat him.

TONY BLAIR REFORMS THE LABOUR PARTY AND TAKES OVER BRITAIN

You'd think the Labour Party would get tired of losing, after four consecutive defeats in British elections from 1978 to 1997. But for two decades, the labor union leaders who ran the party didn't get the message—that the British public didn't like them, and that their continued leadership of the party would assure its continued defeat.

Nor did the leftists, who dominated the party ideologically and insisted on an anti-American foreign policy, understand that the party's advocacy of unilateral disarmament wasn't sitting well with British voters. Worried that the "unilateralists" and labor bosses would ruin the country, swing voters backed the Conservative Party long after its basic ideological positions had lost currency. As Labour historian Keith Laybourn notes, voters came to believe that "the Labour Party was . . . controlled by the trade unions, committed to nationalization, wedded to pacifism, and incapable of dealing with the nation's finances."

Resisting every effort at reform, the coalition of labor unions and leftist unilateralists "learned nothing and forgot nothing"—the description once used for the Bourbon kings of France. Backing leaders who gave lip service to reform but were unable—or unwilling—to break the union and leftist stranglehold, the party went down to one defeat after another.

It wasn't until 1994, when Tony Blair assumed leadership of the Labour Party, that it was turned decisively toward the center, dethroning the union bosses and adopting a policy of giving unions "fairness, not favours." Moving to the right in foreign policy, abandoning the socialist ideology, and supporting legal curbs on union power, Blair made Labour acceptable to Britain. Dismayed—and bored—by nineteen years of Tory rule, British voters turned to him en masse in the elections of 1997.

Blair had to fight long and hard to reform the Labour Party. It is amazing that he succeeded. But once he had performed his alchemy, turning leaden unionism into political gold, he coasted to triumph in the general election. His defeat of the Tories was as easy as his battle against the unions was arduous, and the political momentum he amassed while reforming his own party was enough to propel him to the prime minister's office.

Labour's long years of exile began in 1979, when the party lost after holding power for two terms. Once the party's biggest allies, Britain's trade unions, had become its weightiest burden. "By the 1970s the unions had become an arrogant and destructive force," Geoffrey Wheatcroft wrote in the *Atlantic Monthly* in 1996, and the "connection with Labour was politically damaging for the party."

The labor unions may have been bad, but the leftists were worse. The Labour Party fell under the influence of groups such as the Militant Tendency, a Trotskyite sect advocating public ownership and unilateral nuclear disarmament. "Labour's failure to remove Militant Tendency until the mid-1980s," Laybourn observes, "ensured that the majority of the electorate saw Labour as too dangerous a party to return to office."

Crippled by its own union and leftist supporters, Labour lost power to Margaret Thatcher's Conservatives by 45 percent to 38 percent in 1978. But that was only the beginning of its slow descent into political purgatory.

Not content with losing one election, the unions and leftists proceeded to adopt new procedures for choosing party leaders, solidifying their control over the Labour Party. Their power, and the undemocratic procedures by which it was sustained, led to nineteen years of humiliation and defeat. The system, adopted in 1981, gave the union bosses 40 percent of the vote in choosing party leaders. It allowed them to vote as they pleased, without troubling them to consult their membership. Another 30 percent went to local party leaders, who similarly could support whomever they liked without even asking the party rank and file. The remaining 30 percent of the power went to Labour members of parliament (MPs), who were, at least, elected by the voters.

Flexing its muscles, the newly empowered Left voted to hobble the party with a platform advocating unilateral nuclear disarmament. When Tory prime minister Margaret Thatcher waged a popular war against Argentina over the Falkland Islands, Labour's advocacy of disarmament led many to conclude that the party "lacked a strong sense of patriotic commitment."

When leftist Michael Foot became the party's leader in 1981, it was too much for the moderates. Four relatively conservative Labourites—Roy Jenkins, David Owen, Shirley Williams, and Bill Rodgers—bolted to form a new Social Democratic Party in alliance with the small Liberal Party. Ultimately, twenty-nine of Labour's MPs left to join them, providing a temporary refuge for voters who opposed the Tories but felt Labour had gone too far to the left.

Oblivious to the consequences of the moderate exodus, the unions and the Left continued to run the Labour Party into the ground. Its "manifesto" for the election of 1983, *The New Hope for Britain,* advocated socialism in economic policy, unilateral disarmament, and subservience to the trade unions. MP Gerald Kaufmann called the document "the longest suicide note in history."

Once again, voters showed their discontent with Labour by sharply rejecting it in the 1983 election by 44 to 28 percent—more than double Labour's margin of defeat in the 1979 balloting. One in three of Labour's voters defected to the new Social Democratic–Liberal Alliance which got 26 percent of the vote, and came within two points of passing the Labour Party.

Stung by defeat, Labour then elected a new leader, Neil Kinnock, who wanted to move the party "to the middle ground in order to recapture the support it had lost in the previous two general elections." But the unions and the Left didn't change, because they didn't want to.

No sooner had Kinnock taken over than his efforts to promote moderation were undermined by a coal strike that began on

March 9, 1984. As the walkout turned violent, Thatcher's Tories used the strike to discredit the Labour Party. When Arthur Scargill, the union president, drew immense and visible support at the 1984 Trade Union Conference, the impression spread that the labor movement and the Labour Party embraced his radicalism.

"To deal with these challenges," Laybourn notes, "Kinnock attempted to distance himself, and the Labour Party, from the conflict . . . and refused to commit any future Labour government to giving amnesty to any miner convicted of a serious crime, a commitment requested by the [union]." But he was unable to shake the image of violence that the strike left in its wake.

As if identification with union violence wasn't enough, the Labour Party also reminded voters of its reputation for fiscal irresponsibility when Greater London and two other cities, whose local governments it controlled, refused to comply with spending caps set by Thatcher's national government. "Both the [cities'] non-compliance . . . and the miners' strike . . . left the Labour Party identified with the extreme Labor left, something which Kinnock had tried to avoid."

Kinnock's bid to move the party to the center was further crippled when the 1984 Labour Party Conference decisively rejected his plan to reduce union power by replacing the hopelessly biased process for choosing party leaders with a system based on the inconvenient idea of "one member, one vote." The following year, the Left once again rejected the hapless Kinnock when he tried to drop Labour's support for unilateral nuclear disarmament. The Labour Party Conference instead reaffirmed its endorsement of a non-nuclear defense policy, adding for good measure a demand for the removal of the U.S. nuclear bases that protected Britain.

Kinnock's press secretary, Patricia Hewitt, summarized the party's problem when she wrote that the public still saw it as part of the "loony, Labour left." As the Tories noted in their 1984 campaign guide, "Despite all the rhetoric, the Labour Party remains a wholly-owned subsidiary of the trade union movement."

Frustrated in every attempt to move to the center, Kinnock, Labour, and the loonies lost the 1987 election by 43 to 32 percent, while the Social Democrat–Liberals got 23 percent.

The story of the Labour Party's self-destruction will strike a responsive chord with anyone who has tried to turn around a self-defeating company, organization, or corporation. No matter how obviously suicidal the polices of the leadership may be, it can be downright impossible to get an intransigent body to change direction. Few are ever willing to step up and declare that the emperor has no clothes.

But Kinnock did get one thing right. He discovered a new outspoken moderate Labour MP named Tony Blair, and rapidly promoted him through the party ranks.

Blair had first joined the Labour Party in 1975, and attended his first meeting in 1980. Describing himself as "basically a centrist," he told a newspaper in 1982 that "I want the internal differences in the party to be forgotten, so that we can expose the record of the government and put forward the socialist alternative." But before long Blair would drop the word *socialism*—and its attendant principles—from his vocabulary.

From his first days in the party, Blair sided with those who backed the one-member, one-vote reform. He noted that "the Left's position is often inconsistent on democracy. It will advocate party democracy, yet refuse one member, one vote. . . . It will talk of decentralisation yet find itself at a bizarre and remote distance from most of the opinions of those to whom 'power' is supposed to be given."

Concerned that Blair was too moderate, left-wing MP Dennis Skinner accused him of "betraying socialist principles." But socialist or not, he got the Labour nomination for Parliament, and survived Labour's electoral massacre in 1983 to win a seat representing the constituency of Sedgefield in the House of Commons.

The massive Labour defeat led to a change in Blair's attitude. Before the election, Blair had still believed that Labour could win without changing its positions; now he realized that "the Labour Party needs to re-create itself." His "position [after 1983 was] much more destructive of the Labour Party as we know it. . . . He came to believe at that period . . . [that] without fundamental change in the Labour Party, we would never win."

On *BBC Newsnight* in June 1983, Blair asserted that "the image of the Labour Party has got be more dynamic, more modern. . . . Over 50 percent of the population are owner-occupiers—that means a change in attitude that we've got to catch up to."

From the beginning, though, Blair grounded his commitment to change in a simple realization—that party reform was the key to electoral success. Intuitively, he seemed to grasp that if the battle to reform succeeded, it would lead him to the prime minister's office. If it failed, it would mean political death.

Stressing the need for internal change in Labour's policies and procedures, Blair envied Margaret Thatcher's success in transforming the traditional, aristocratic, hidebound Conservative Party into a zealous group of free-market activists. Blair wrote that Labour needed "profound changes in ideas and organisation," just as "the key to Mrs. Thatcher's political success has been in destroying and re-creating contours of electoral support."

Anxious to encourage reformers, called "modernizers," Kinnock invited Blair to sit on the front opposition bench in Parliament, giving him the position first of Energy and then of Employment Minister in the shadow cabinet—the opposition party leaders primed to form the real cabinet if their party should take power.

But the promotion was a decidedly mixed blessing. As Employment Minister, Blair became the point man for the Labour Party's response to Thatcher's bold program of labor union reform. Knowing that "Labour's links with the unions were the source of the Party's greatest strength—and weakness," in the words of biographer John Sopel, Blair had to steer a ticklish course to appease the public desire for regulation and reform without alienating the party's union base of support.

Thatcher's measures to rein in union power and curb the unfettered right to strike were popular with the voters, but hated by the unions. Thatcher demanded that a majority of rank-and-file union members had to approve such a strike before it could be called. She passed laws regulating how and where strikers could picket. Union workers were no longer allowed to receive amnesty for crimes committed during protests or strikes, and the courts were empowered to intervene and sequester union funds in reaction to illegal strikes.

But the law that catalyzed the most union opposition was the Conservative government's insistence on ending the "closed shop," the practice of negotiating contracts that required all employees to be union members. Trade unions had fought hard for the closed shop for decades, and the sudden reversal of their cherished winnings caused howls of anguish. Indeed, unions disputed that the government had any right at all to legislate labor practices.

But both Kinnock and Blair understood that to oppose Thatcher's reforms would destroy their effort to modernize the Labour Party. Voters would lose any faith they might have had in a "new" Labour Party, and conclude that it was controlled as surely as ever by the same old union bosses.

It was Tony Blair's fight to win, and he did: He got the Labour Party to endorse an end to the closed shop, along with most of Thatcher's other reforms. He accomplished what most felt was impossible. And he did so without alienating the bulk of the union leadership or membership. It was the first virtuoso performance of a master politician, a successful effort to reform his party while retaining its support.

The key to Blair's success was his recognition that he could only win by genuinely convincing most union and party members of the need to change. Although Blair knew he might be able to jam the policy change down their throats, he also recognized that doing so would cause a schism, weakening the party and its image without achieving his goals.

Reformers usually find it easy to villainize their institutional opponents when their own followers are standing before them cheering and throwing bouquets. But it's quite another thing to go into the teeth of the enemy and sit down with him to explain why reform is needed. The impassioned reformer's signal mistake is often to give the opposition a wide berth, and spend most of his time with the good guys.

John Monks, at this point the deputy general secretary of the Trade Union Congress, described how under Neil Kinnock's party leadership "there were times when the unions were made to feel like an ugly Siamese twin brother, whom the Labour Party would dearly like to get rid of, but couldn't."

Blair was different. As the Shadow Cabinet member for Employment, he flung himself into the task of selling union leaders and members on the need for reform with incredible dedication and energy. He regularly met with union members and had "an open-door policy" for union leaders, Sopel notes. "All the union barons (with perhaps the odd exception) were impressed at the speed with which Blair put himself around. They also knew he spoke with the authority of the Leader's office, and were strangely flattered that one of Labour's rising stars had been put in charge of a brief that had often been used as a political dumping ground. . . . Blair would be available and visible. . . . He made a point of dropping in on all the senior union leaders, and they found him intelligent, friendly, keen, ambitious but absolutely determined about exactly where he wanted to take the Labour Party."

Blair's message was clear. Regulation of unions by the government was inevitable. The only option labor had was to make the rules workable. As he put it: "The issue today," he said, "is not 'law or no law' but 'fair or unfair law.'"

Blair's greatest challenge was to confront the unions with the unwelcome truth that the closed shop was a thing of the past. "For the ordinary union member," Sopel notes, "it seemed that Blair was attacking their very way of life. But, in an extraordinary sequence of events, Blair, through a mixture of opportunism and skill, persuaded the unions in a matter of days to agree to give up something they had fought for over decades."

Blair presented the union leaders with a tight, closely reasoned case for reform. He made it clear that unless the Labour Party was able to change its image of union domination, it could never again achieve power in Britain. The closed shop was widely unpopular, he explained, and Labour had to jettison it from the party platform if it was ever to have a chance to rise again. Facing the bitter alternative of another decade without power, the union leaders yielded, grudgingly, to reality and accepted the need to change.

Just as important as the policy victory was the measured tone that accompanied it. When Blair achieved his aim and secured the Labour Party's backing for an end to the outdated closed shop, he took great pains not to dance on the unions' grave. Rather than making a fuss over the undeniably major change, Blair was content to issue a simple and quiet "policy clarification," couching the shift in terms of its reaffirmation of the rights and interests of the individual. By the time Blair was finished, he had turned the closed shop issue around against Thatcher.

An old proverb suggests that Irish diplomacy is "telling a man to

go to hell in such a way as to make him look forward to the trip."
Blair's pitch to the unions would amply meet this definition.

The lesson for anyone who would attempt to change an organization's direction is obvious. One-on-one persuasion is the key. The
reformer has to sit down, face to face, with those whose wings he
would clip and explain why it is necessary and how they can survive
under the new system. Nobody can reform an organization long distance. It takes hands-on leadership.

Blair's charm and persuasion worked, in part, because unions in
Britain were a lot weaker in 1990 than they had been in 1980. The
arrogant labor leaders of 1980 had 6.5 million members at their beck
and call. By 1990 their ranks had fallen to 4.9 million.

The British unions had also had quite enough of Margaret
Thatcher and the Conservatives. The experience of watching
Thatcher's policies strip them of one-quarter of their members—and
dues—reminded the unions that there was no substitute for a
Labour victory. Blair might not have been the compliant tool that
Labour prime ministers had been in the past, but at least he had the
political skill to marshal popular support for his reforms—skill that
held enormous promise as Labour looked forward to new elections.

As the 1990s dawned, Blair's push toward moderation was taking
hold. Labour made progress in municipal elections and even gained
an edge over the Conservatives in the race for the European Parliament. By the election of 1992, the "new" Labour Party was ready for
its public debut.

Under Blair's guidance, the party declared that it would retain
most of the Conservative government's union legislation, especially
the ban on closed shops. Thus Blair succeeded in dismantling nearly
every principle the Labour Party had once held dear.

Yet while the party's policies had moderated, the old leftists were
still in charge. Three-quarters of Labour politicians favored more
nationalization of private industry, while only one-quarter of the voters agreed. A majority of Labour's candidates for office backed elimination of Britain's stockpile of nuclear weapons, but only 14 percent
of the voters agreed. And every single Labour candidate polled was
in favor of higher government spending—but only 57 percent of the
voters went along.

Despite all the progress he had made in modernizing his party,

Kinnock and Labour lost yet another election in 1992. Neil Kinnock simply was not up to the job of defeating John Major, Thatcher's heir as Conservative leader. As Laybourn noted, one report asserted that "Mr. Major did not win the election. Mr. Kinnock lost it"; another wrote that "Voters just did not believe Mr. Kinnock was fit to run Britain."

Blair himself recognized other, more fundamental reasons for Kinnock's defeat. "The worry of the electorate in 1992 was not that Labour had changed, but the concern was that the change was superficial. . . . The reason Labour lost in 1992, as for the previous three elections, is not complex, it is simple: society had changed and we did not change sufficiently with it. . . . Labour does need a clear identity based on principle, not a series of adjustments with each successive electoral defeat."

The real prescience of Blair's analysis is not that he recognized the need for party reform, but that he realized that reformation was the key to electoral success. The Tories had nothing to do with Labour's defeat. The party was losing because of self-inflicted wounds.

Defeated by 43 to 35 percent, Kinnock resigned, and in July 1992, John Smith became the new Labour leader. But Smith's commitment to reform was unclear. Asked whether he thought that Smith was a modernizer, Blair answered candidly: "I just wish I knew."

His skepticism may have been warranted, but it glossed over Smith's one undeniable accomplishment: During his brief tenure, Smith opened the way for Blair's succession to the party leadership by finally reforming the system for choosing party leaders, adopting a modified one-man, one-vote rule. Smith replaced the 40–30–30 division of votes among unions, local parties, and MPs with an even three-way split. But, more important, he changed the system that empowered union leaders and local party constituency chairmen to decide, unilaterally, for whom to cast their organization's votes. After Smith's and Blair's reforms, they would actually have to hold a vote of their members and divide their organization's votes proportionately according to the vote. It was the dawn of a new era.

In reforming a political party or any institution, the key is not only what you do change but what you take care not to change.

Even as Smith and Blair curbed the power of the unions and local political leaders, though, they were careful to leave them enough power to keep them in the Labour Party. After all, even though union leaders had to consult with their members, and party bosses with their rank and file, the unions and local party constituency groups still each cast one-third of the votes for party leader.

Blair and Smith also took care not to step on the toes of the Labour MPs. Indeed, the reform left them with slightly more power, since their share of the votes for party leader rose to one-third from 30 percent. Politics is, after all, a game of inches.

When John Smith died, on May 12, 1994, Blair's accession to power was finally at hand. With Smith gone and the one-man, one-vote system in place, it was clear that a "modernizer" would take over as leader. For a moment there was some question whether Blair or fellow reformer Gordon Brown would become the new party leader; when the latter withdrew, Blair's election became a foregone conclusion.

In his shoes, many would have coasted to an easy, uncontested victory as party leader. But once again Tony Blair's keen political instincts led him in a counterintuitive—but highly beneficial—direction. Blair sensed that 10 Downing Street might at last be within reach of his party. But he also realized that it would take a dramatic spectacle to attract the nation to his cause. If he could reform his party in full view of the nation, he perceived he would be so empowered that he could capture the ultimate prize. What Blair needed was an open fight, closely followed by the British press and public. He had to slay the dragon with fanfare and showmanship, to convince the country that he was the real reformer he promised to be.

Blair had the support, the funds, the platform, and the opportunity to win. The only thing he lacked was an opponent. Blair met with fellow Labour MP John Prescott to discuss whether he should just accede to party leadership or whether he should run in a contest. Fortunately, Prescott was thinking along the same lines as Blair, and advised that a fight would be helpful in demonstrating how the party had really changed. Prescott offered to be the sacrificial lamb and run against Blair.

With Prescott as an opponent, Blair announced his candidacy for Labour leader on June 11, 1994. "It is now time to complete the journey started by Neil Kinnock and John Smith," he declared, "and move from the politics of protest to the politics of government. To end the long years of oppression—dark years for us and often for our country and to seek the trust of the British people in shaping their future. . . . We must state a new vision of our country, a vision of hope that the world as it is, is not the world as it is meant to be."

Conscious that he was auditioning for the British public, Blair laid out a series of positions, each in sharp contrast to the traditional views of Labour politicians of earlier days. To business, Blair admitted that Thatcher's "new right had struck a chord. There was a perception," he recalled, "that there was too much collective power, too much bureaucracy, too much state intervention and too many vested interests created around it." Blair urged Britons to look ahead: "The task is not to return to the past. The era of corporatist state intervention is over, the task is to move forward by renewing forms of economic and social partnership and co-operation for the modern world."

To those who fretted that Labour would raise taxes, Blair responded, linking a "high tax economy" with a "low success economy."

Taking a page from the Bill Clinton handbook, Blair made education reform a central element of his campaign, reversing Labour's traditional opposition to evaluation and testing. Calling for the " 'sacking [of] incompetent teachers,' he drew fire from the second-largest teachers' union."

Blair took special pains to distance himself from labor unions. "They are not going to be shut out in the cold or told they are not part of our society," Blair said. "They are an important part of the democratic process. But we are not running the next Labour Government for anything other than the people of the country."

Echoing Christ's admonition to "render unto Caesar the things which are Caesar's and unto God the things which are God's," Blair said: "Unions should do the job of trade unions. The Labour Party must do the job of the government. The British people expect Labour to govern for the whole country, and we will."

Beyond his centrist positions, biographer John Rentoul notes, Blair "talked in phrases more familiar to the Right of politics, in a moral language that was utterly unfamiliar to most of the Labour Party that elected him." Instead of class warfare, he spoke of family values. Rather than discuss union militancy, he focused on issues like education and crime. In vocabulary as well as in issues, he dragged the Labour Party into the 1990s.

That Blair's centrism played well with the British public was obvious. But, surprisingly, it also worked within the Labour Party. On
July 21, 1994, he became party leader, winning 57 percent of the vote.
As expected, he got 61 percent of the MPs and 58 percent of the local
members—but, amazingly, he also won a majority among labor
union members. It was the greatest mandate any Labour Party leader
had ever enjoyed. As Sopel notes, though, "only a tiny handful of
people had cottoned on to just how radical Blair intended to be"
once he got into office. As Geoffrey Wheatcroft noted, "a party has
been captured from the inside, and by a man who in his heart
despises most of that party's traditions and cherished beliefs . . . It
might be that the Labour Party as it has existed for nearly a hundred
years has indeed disappeared."

His arranged opponent, John Prescott, joined the chorus of
praise. "This man, our new leader, has got what it takes. He commands moral authority and political respect. He has the energy and
vitality to win people over to Labour. . . . And he scares the life out of
the Tories."

Blair promised that he wouldn't stop until the "destinies of our
people and our party are joined together again, in victory at the next
General Election."

Calling his party "New Labour," Blair said, "I will tell you how it
works. Not through some dry academic theory or student gospel of
Marxism. It works when every person who wants to, can get up in the
morning with a job to look forward to, and prospects upon which to
raise a family."

In Kennedyesque manner, he called upon the young people of
Britain to "join us in this crusade for change. Join us. Of course, the
world can't be put to rights overnight. Of course, we must avoid foolish illusions and false promises. But there is, amongst all the hard
choices and uneasy compromises that politics forces upon us, a spirit
of progress throughout the ages, with which we keep faith." He
ended his statement eloquently, pledging to lead Labour "through
courage and compassion and intelligence, but most of all through
hope—the small, broken moments of hope that forever are worth an
eternity of dull despair."

His message and his strategy were working. In the days after Blair's
election, Labour soared to a twenty-point lead over the Tories in

national polls, with particular gains among the middle class. But Blair had one more hurdle to surmount before he could fully convince the people that his pledges to change were genuine. He faced the daunting task of reforming the infamous Clause IV of the Labour Party constitution, which committed it to common ownership of the means of production—a politically inconvenient reminder that Labour was, after all, a socialist party. Written in 1918, as revolution swept through Russia, the language was pure Karl Marx: "To secure for the worker by hand or by brain the full fruits of their industry and the most equitable distribution thereof that may be possible upon the basis of the common ownership of the means of production, distribution, and exchange."

There was every reason to expect that the trade unions would oppose the repeal of Clause IV. But then Blair worked his magic. He made personal appeals to key union heads and Labour Party members, and secured the endorsement of his friendly adversary John Prescott. But Robin Cook, then the trade and industry spokesman, frustrated Blair by blocking his proposals at the party conference.

Blair and Prescott, determined to prevail, drafted an alternative to Clause IV accepting "a competitive market economy with a strong individual and wealth generating base." Rejecting common ownership as a "reflex answer to all market failures," the new language promised, "We do not believe in absolute arithmetic equality."

At the close of 1994, Blair personally addressed thirty thousand party members and asked them to support his new language. The gambit paid off: At a special party conference on April 29, 1995, Blair's new Clause IV passed with just under two-thirds of the vote. "In the end, the majority of trade unions and constituency parties dared not vote against Blair if they wished the Party to have a significant political future."

As always, Blair was careful to be a generous victor, making a host of smaller strategic concessions to placate the left wing of his party. But his major accomplishment was clear. As he later reflected, scrapping Clause IV "showed me what I intuitively thought but wasn't sure of: that the party was actually behind change."

Why did the labor unions stick with Blair? Why did they put up with the dilution of their power and accept the leadership of a man seemingly determined to diminish it further? There is a simple answer: Margaret Thatcher. Her Tory party's attack on the unions during the 1980s had stunned labor leadership, sweeping away a century's worth of labor advances—the closed shop, the right to strike without consulting membership, freedom from government regulation, the right to picket when and where they chose, and virtual immunity from judicial oversight. Lingering resentment over Thatcherism was the adrenaline driving Blair's victory. To the reeling Labour Party, if the alternative was Thatcherism, they'd stick with Blair.

Labour's decapitation of the unions did what it was meant to do: as the party cut the power of the unions that had been its core, it gained public confidence. A Gallup poll conducted soon after Blair became leader showed that 53 percent of Conservative voters would be happy or content about a Labour government. An *Economist* poll in October 1994 showed that "two thirds of the electorate now agree that Labour has changed for the better. Its values are widely felt to reflect their own values, and they think that the party understands the 'needs of people like me.' "

Blair had turned the tide; victory was his for the taking. The general election of 1997 was over before it started. Blair had defeated old Labour, and there was nothing the Tories could do to defeat new Labour.

The irony of Blair's procession to the prime minister's office was how easily the Tories folded. Once Blair had conquered his own party's left, he quickly gained standing and confidence among the British people. His internal reform of his own party boosted him to a lead in national polls that he never relinquished. One preelection poll showed that 77 percent of the British public found him to be a "strong leader," while only 35 percent felt that way about Tory leader John Major.

Blair's triumph also highlighted the potential for overlap between the reform strategy and triangulation: Blair proceeded to steal the Conservatives' thunder on crime, health care, taxes, free-market economics, union regulation, and seemingly every other topic that mattered to Britons.

Hobbled by having imposed a tax increase it had promised to avoid, the Conservative Party lost the confidence of the British people at about the same time that Blair was gaining it. The tax issue was a priceless weapon in Blair's arsenal: "If we can force a draw on tax," he said, "we will win." Labour's television advertisements attacked the Conservatives, charging that they had passed "twenty-two tax increases since 1992" and warning that the Conservative pledge not to raise taxes meant nothing since the Tories had broken such promises before and might very well do so again.

Adopting American campaign techniques—including rapid response, negative ads, spin doctors, sound bites, and media events—Blair ripped Major as weak. His campaign goals focused on education, crime, health care, jobs, and taxes. Change the faces, the accent, the vocabulary, and the venue, and it could have been Bill Clinton talking.

While the Tories tried to rally old fears by warning of "New Labour, New danger," they knew from the beginning of the campaign that they likely couldn't win. Major's chancellor of the exchequer moaned "bad luck will be engraved on our tombstone," and another Downing Street aide suggested, "The Conservatives have lost the most precious commodity in politics—the trust which persuades voters to give their politicians the benefit of the doubt."

Blair took the high road: "The election will be a battle between hope and fear. People will be saying Labour is going to do this to you and to do that to you. We have got to settle and reassure the people."

By now, though, Blair had so thoroughly won the public's confidence that when the *Daily Telegraph* and *Daily Mail* ran front-page stories on rumored new Labour plans to boost union powers, Blair succeeded in dismissing the charge as "complete nonsense." He said: "I did not spend three years turning the Labour Party into a modern party that is true to the principles of progress and justice . . . to go backwards."

On election eve, Tony Blair went to his home constituency in Sedgefield, and urged people there to vote Labour. Otherwise, he warned, they faced another five years of the "most discredited and sleazy government." Change was only hours away, he promised: "It is twenty-four hours to save our [National Health Service,] twenty-four hours to give our children the education they need, twenty-four hours to give hope to our young people, security to our elderly."

The *Guardian* called it a "triumph." The *Daily Mail* said it was a "massacre." The *Express* and *Daily Telegraph* called it a "landslide." On May 2, 1997, the British public overwhelmingly voted for Labour

Party candidates in the general election. The Labour Party beat the Tories 43 to 31 percent, winning a huge majority of 418 seats in the House of Commons to the Conservatives' 165.

Middle-class homeowners—the group militant Labour once dismissed as the "petit bourgeoisie"—voted for Labour by 41 to 35 percent, underpinning the national landslide.

Exit polls confirmed that Labour had seized the public trust away from the Tories. Voters trusted Labour over the Conservatives to help schools, to reduce corruption, and to improve the economy. Even on the tax issue, they trusted Labour.

As Blair confidant Peter Mandelson said: "There was no reason left not to trust Labour. All the old 'ifs,' 'buts,' and 'maybes' had gone. We removed the target. Without New Labour the Conservatives could have won again."

The lesson of the Blair triumph is a heady one for anyone looking to turn reform into victory: If you can beat your domestic adversaries—win the inside game—and keep the loyalty of those you have bested, the people will reward your efforts.

Halfway around the world, Japan's Junichiro Koizumi pulled off the same trick in 2001. He reformed his own party, the venerable Liberal Democratic Party (LDP), which had ruled since the 1950s, and in the process became a national hero.

KOIZUMI REFORMS JAPAN'S RULING PARTY . . . AND TRANSFORMS JAPANESE POLITICS

The Liberal Democratic Party (LDP), which has dominated Japan's political landscape for decades, is neither liberal, nor democratic. Nor is it a party.

It is, essentially, an ongoing political deal, constantly renegotiated to preserve a delicate balance of power among its factions, and within them, as the system's players gain or lose power. With no clear or consistent guiding philosophy or principles, the LDP has one raison d'être: to represent the economic special interests that rule contemporary Japan. Together an array of bankers, construction contractors, drug manufacturers, farmers, car companies, doctors, hospitals, and dozens of other potent commercial interests exploit the Japanese people and together they dominate its major political party.

The LDP not only controls the Diet (parliament), but also dominates the civil service bureaucracy whose regulations shape every aspect of Japanese life. There is no real distinction between career bureaucrats in appointed positions and the elected officials or cabinet members of the LDP. Indeed, when the bureaucrats (who often rule with arbitrary—if sometimes petty—tyranny) retire from their government jobs, they often run as LDP candidates for the Diet.

The LDP politicians coexist happily alongside the bureaucrats, each bolstering the supremacy of the other. The bureaucrats run the country, micromanaging each aspect of a largely closed economy, while the politicians dole out government goodies among their special-interest clients. Highway contracts, special-interest laws, tax exemptions, open competition limitations, import curbs, and other species of largesse flow from the politicians in return for the campaign contributions that sustain them.

This world of financial incest, vice, and corruption mattered little to ordinary Japanese people as long as the nation's economy boomed and jobs proliferated. But, in the 1990s, as the flaws in this rigidly controlled economy were exposed, the need for reform became apparent—to all but those in power.

During the last decade voters chafed under the LDP-bureaucrat alliance, rallying first to one candidate and then to another as each pledged to back reform. Alienated by years of broken promises, voters soon began to despair over the inability of the system to reform itself.

Then came Junichiro Koizumi, an LDP politician who had spent his career in relative obscurity, climbing to the top of one of the party's many factions. Koizumi was different. As no previous LDP leader ever had, he grasped the need for fundamental reform—and, like a Japanese Gorbachev, he challenged the party's elders from within.

Koizumi burst onto the Japanese landscape at the beginning of the twenty-first century like a rock star. Voters who had been bored to distraction by the elderly, staid political leaders their nation had endured for decades, thrilled to Koizumi. Used to slow-moving prime ministers who stumbled in and out of power in dizzying succession, they saw in Koizumi a new style of leader with a real commitment to reform. Young, handsome, energetic, humorous, modern, stylish, and irreverent, Koizumi took on the political power brokers as he smashed to the top of the party machinery. One opposition leader told me he was "a man totally without fear."

In a series of hard-fought political battles, he challenged his own party, pulled it back from the verge of extinction, and led it to a smashing victory in the Upper House elections of 2001.

As this is written, the ultimate fate of Koizumi is unclear. Japanese politics is far too mercurial to place any long-term bets. But the story of how Koizumi came, seemingly out of nowhere, reformed his party, and used the momentum of his early victories to take over Japan, deserves attention.

When Tony Blair took control of the Labour Party, it had suffered four consecutive election defeats over the preceding sixteen years. Even the dunderheads who ran Britain's unions had realized by then that they weren't exactly awash in popularity. But Koizumi took over an LDP that was still undefeated, still in power. Though it had been forced since 1993 to govern in coalition with the cultlike Buddhist Komei Party, the LDP had not lost its grip on the Japanese political system. Yet Koizumi, unlike Blair, wasn't even forced to go through the long, elaborate courtship of the party elders.

How did Koizumi do it?

The LDP had been losing support, bit by bit, throughout Japan's decade-long recession of the 1990s. Its vote share had slipped from 33 percent in the Upper House elections of 1992, to 27 percent in 1995, to 25 percent in 1998. Many political observers wondered whether 2001-2002 would be the years in which the LDP finally lost power.

In a desperate attempt to revive the economy and stave off defeat, LDP governments had adopted successive fiscal stimulus packages totaling almost $1 trillion (100 trillion yen), creating a huge deficit but little growth. The nation's banks were in equally desperate condition, having issued a fortune in bad loans during the highly inflated real estate market of the 1980s.

When faced with a similar crisis in its savings and loan associations in the 1980s, the United States government ruthlessly closed defunct banks and prosecuted those who had run them. But the LDP would not, and could not, turn on its allies in the banking sector. Determined to prop up the banks and the businesses whose bad debts burdened their ledgers, the Japanese government was forever postponing the day of reckoning. The resulting crisis handcuffed both government and banks, leaving the economy to fall even further each month. Unemployment rose. Suicides increased. And the LDP's popularity nose-dived.

Soon, Japan's fledgling opposition parties—principally the Democratic Party of Japan (DPJ)—began gaining support. The DPJ, a loose federation of reformers, former socialists, and disenchanted former LDP members, was committed to a far-reaching program of change. With its allies in the Liberal and Socialist parties, the DPJ found its vote share moving up in each election as the LDP wore out its welcome with the Japanese people.

For the LDP, prime minister after prime minister came and went, each lasting barely a year, until the party finally hit bottom by naming Yoshiro Mori as prime minister on April 5, 2000. Karl Marx said that history repeats itself, the first time as tragedy, the second time as farce—and he must have been anticipating Mori when he said so. As Mori's administration unfolded, his ineptitude and inability to cope with modern Japan were so apparent that nobody could doubt the need for change. If Koizumi was the Gorbachev who saw the need for change from within, then Mori was the equivalent of Gorbachev's predecessor, Konstantin Chernenko: the last feeble reminder of what was wrong with the system.

Chosen as party leader in "a secret deal among several [party] heavyweights," from the start of his administration Mori was so tainted by the closed process that selected him that he had very little credibility and only a slender margin of support. And his personal gaffes only eroded what little backing he may have enjoyed. Shortly after he took office, Mori called Japan a "country of gods centering on the Emperor," a statement that reminded many Japanese of the mystical musings of prewar right-wing militarists. Soon thereafter, he saw his disapproval rating reach 70 percent.

Corruption also drained Mori's political capital. Masakuni Murakami, a former LDP labor minister who had played a key role in Mori's selection, was arrested on March 1, 2001, for allegedly receiving $700,000 to promote a public works project—a development the *Japan Times* called "a serious additional blow to the already faltering Mori administration." (Murakami denied the charges.)

The economy, which was just recovering from the two long recessions of the 1990s, now headed into a third one. The Nikkei index of the Tokyo stock exchange fell 40 percent in the year after Mori took office.

But the worst blow came on February 9, 2001, when Mori learned that a United States naval submarine and a small Japanese fishing boat had collided in the Pacific. Nine Japanese, including four students, were killed. Already angered by a series of sexual assaults by

U.S. soldiers based in Okinawa, the Japanese were furious with the American navy. But when they learned that Mori had refused to let news of the disaster interfere with his golf game—reportedly playing for two additional hours after learning of the tragedy—they very quickly turned their rage toward their own prime minister.

With an election for the Upper House scheduled for July 29, 2001, and Mori's disapproval rate soaring to 82 percent, the party elders realized that the time had come to cut Mori loose. All of Japan was watching to see when the ax would fall. Finally, on March 10, 2001, Mori announced that he would convene a meeting of the party in April to select a new leader.

Everybody expected the selection process to follow the usual pattern—a closed session, followed by a fait accompli. The *Japan Times* commented that "the indications are that Mori's successor will be chosen on the basis of the factional balance of power in the LDP, as was Mori last April. The LDP is making little effort to reform itself and is inviting increasing public distrust." Once again, it seemed, the various factions of the LDP would cut a deal among themselves, and then submit their choice of a new leader to the LDP members of the Diet for rubber-stamp ratification.

Yet in this newly volatile political climate, the old ways would no longer prevail. Now the party's legislators and regional representatives started to call for a full-scale election in which all party members would be eligible to vote. DPJ leader Yukio Hatoyama joined the chorus of criticism over the LDP's traditionally closed selection process. "The way a prime minister is chosen and then dumped all behind closed doors is a complete denial of parliamentary democracy," he commented in March.

Obtuse though he was, even Mori himself began to sense the mounting criticism of the traditional selection process. "Now we have to start from scratch," he told a Tokyo party meeting, "and make efforts to win the trust of the people." And he proposed the radical idea that called for the new LDP leader to be chosen "in a way that allows various opinions to be reflected."

Outside the meeting, "a group of 40 LDP members of the Tokyo Metropolitan Assembly showed up in front of [the convention] and started handing out leaflets harshly criticizing the party leadership as thousands of rank-and-file party members" gathered around them.

"Wearing white headbands calling for the 'regeneration of the LDP,'" the *Japan Times* reported, "the rebellious group defied orders from party leaders to stop their demonstration."

Demanding that the party choose Mori's successor through an election open to all party members, they warned that if the process remained unchanged, the LDP would face "the most tragic and worst scenario—that is, death as a political party."

Facing an outpouring of pressure from the grass roots, party leaders decided to grant ordinary party members a small share of the 487 delegates to the party meeting that would decide who would lead the LDP. Their compromise granted the rank and file 141 votes, or 29 percent of the electorate, while members of the Diet retained 346, or 71 percent of the vote. While the procedure allowed for token grass-roots participation, party leaders were confident they could still control the process, so absolute was their domination of the Diet members.

Most observers were pessimistic about the selection process, reformed as it was. Political commentator Ryurichiro Hosokawa wrote in the *Japan Times*. "The LDP seems dead, incapable of finding the next prime minister among its leaders." Another commentator was equally doubtful. "For the LDP to be born again, it must reform itself in an open and transparent manner and develop younger leaders by putting them in responsible positions. For that, the party must have a competent leader capable of pushing internal reforms beyond the pale of factionalism. The question is whether the next LDP president will have that kind of ability."

As if to prove the pessimists right, the front-runner for the position of prime minister and party leader was former prime minister Ryutaro Hashimoto, whose faction was by far the largest in the LDP, with 102 of the party's 346-man Diet delegation. When party general secretary Hiromu Nonaka, also a member of the Hashimoto faction, withdrew from the race and the group unified behind Hashimoto, the veteran party warhorse—whose allergy to reform was well known—seemed the likely winner.

Unconcerned about the new voting rules, the Hashimoto camp was counting on the support of large industry groups, which controlled approximately two thirds of the 2.4 million registered LDP members. But lurking beneath the surface of support for Hashimoto was the first serious threat to conventional party leadership in decades—Junichiro Koizumi.

Koizumi, who had lost to Hashimoto in a race for party leader in 1995 and then again in 1998, was pondering a third run for the post. As the nominal leader of Prime Minister Mori's sixty-one-member faction in the Diet, Koizumi had to avoid the appearance that he was coveting his boss's position. Asked about his ambitions, he said blandly, "All I can do is to support Mr. Mori so that he can fulfill his duties."

When Mori announced he would step aside, however, the way was clear, and Koizumi began discussing his plans in public. "I want to explore ways to respond to the people's expectations by demonstrating resolve to restart the party from scratch."

Koizumi had already enjoyed a long career in Japanese politics. A graduate of the prestigious Keio University, he had been a graduate student at the London School of Economics when his father, a member of the Diet, died in 1969. He returned to Japan to run for his father's seat, losing by four thousand votes, but won a seat in 1972—a position he has held ever since. Moving up through the ranks, Koizumi served as health minister in 1988, and again from 1996 to 1997. Beyond a personal reputation for idiosyncrasy, there was little in Koizumi's past record to indicate that he would be a major force for party reform. But in the thirty-day campaign he was about to launch, all of that would change.

At first, Koizumi didn't appear to have much chance of victory. Coming off two consecutive defeats, Koizumi had only limited popularity among party leaders. Yet he was the only major party figure who truly understood the potential for change implicit in the new selection process. The rank-and-file election, Koizumi perceived, was tantamount to a national primary. And it called for campaign tactics more typical of an American than a Japanese campaign.

It was Koizumi's genius to look beyond conventional wisdom, to discover that there was an opening for an iconoclastic, reformist leader who could steal a march on the party elders and force them to accept reform . . . and to vote for him.

As he launched his campaign, observers like Kuniko Inoguchi, a professor at Sophia University, compared his rock-star hair and youthful good looks to John F. Kennedy. "He resembles JFK very much in that he is the first one to understand the impact of TV and mass media in a democracy, and he is very good at using them. . . . People identify with him and find hopes for the future when times are very difficult. You have to be a very good master of images and symbols, but like Kennedy, Koizumi is convincing the people that it is a difficult time but there are solutions." When others tried to tar him as an oddball, Koizumi answered: "So I'm an eccentric? People judged eccentric in [the Diet] are thought pretty normal by the general public."

But behind the brash appearance of this fifty-nine-year-old man— whose hairdo made him appear much younger—lay a serious commitment to reform—a commitment he underscored by resigning his position as leader of the Mori faction at the very start of his campaign. Factions in the LDP are no loose groupings of like-minded individuals. Rigidly defined and disciplined, the LDP factions are more like separate parties who vie with one another for a larger share of ministerial portfolios and jobs. Condemning the faction system, Koizumi charged that it contributed to the corruption of Japanese politics. He even called for stripping party leaders of a role in the selection of party candidates. "I'd like to see people directly vote for the Prime Minister, like the system practiced in the United States. It can be done, by changing the Constitution. But there's strong opposition in the ruling parties."

Vowing "reform with no sacred cows," Koizumi promised to ignore the demands of the special-interest groups that had defined the LDP in the past. Chief among these was the construction industry, which had lobbied successfully for massive overspending on public works and highway construction. Promising to limit spending on such projects, Koizumi proposed the heretical step of diverting gasoline and car tax revenues earmarked for highway construction to pay for spending on other programs or even to allow tax cuts.

Calling Japan "addicted to debts," Koizumi noted that "budgets at both the central and local levels are out of control, with their deficit adding up to 10 percent of GDP. No modern developed nation has run such deficits before." One American investor calculated that "the Japanese government borrowed at a rate of $40 million an hour, twenty four hours a day." Koizumi pledged to cap government borrowing, tighten public spending, and clean up the banking system.

Koizumi also wanted to force banks to write off bad debts and drive deadbeat companies into bankruptcy, even though he admitted that this would cause "pain." That may have been putting it mildly: The *Economist* predicted that such a step would put "hundreds of thousands, perhaps even millions, of people out of work."

In an economy traditionally averse to job layoffs, Koizumi also proposed to encourage restructuring by granting jobless benefits—unknown in Japan—equal to full pay for one year.

Stirring up a huge fuss, Koizumi proposed to privatize the postal savings system in which, *Business Week* reported, the "millions of Japanese have stashed $2 trillion in deposits." For decades, the money had been "shoveled into wasteful government projects, to Koizumi's chagrin." The magazine noted that "of all his [Koizumi's] reform proposals, none fires him up like postal savings privatization. Japanese bankers, who view the system as a tax subsidized monster that diverts deposits away from the financial markets, are amazed at Koizumi's intensity. . . . Koizumi would love to break up the postal savings system into regional savings funds, then eventually privatize them as individual banks." Koizumi's attack on the postal system, struck at the very heart of the LDP power: Local postal chiefs and their families were an important part of the party's voting base.

As he fleshed out his reform proposals, it was increasingly evident that Koizumi meant what he said when he pledged reform "without sacred cows." Economists predicted that Koizumi's plans would "hike Japan's 4.7 percent jobless rate (a height unimaginable a decade earlier) to 6 percent, and 1.5 percentage points could be shaved off annual growth for the next two years." Koizumi remained firm. Without "resolutely implementing structural reform," he said, "we won't get a recovery."

Could Koizumi stay in the LDP while advocating such drastic reforms? He felt he could, he said, because "the time will come when people realize that it's necessary. Japan is facing a crisis—the country keeps borrowing money to service its debts—but neither the public nor the politicians have a sense of crisis. When there was an oil crisis, it was said that Japan would suffer the worst damage. So Japan borrowed money and expanded its economy. But the government did not pay back its debts when the economy became strong. And it kept borrowing until the country was stuck with huge debts. People don't understand that privatizing the postal system doesn't just affect the post and telecom sector; it affects the entire gamut of administrative and fiscal problems."

On foreign affairs Koizumi was no less revolutionary, hinting that he wanted to amend Article 9 of the U.S.-imposed Japanese constitution, which forbids Japan to have any military forces other than for self-defense. Rejecting calls to change Japanese textbooks to depict Japan's wartime atrocities more accurately, Koizumi proposed to—and eventually did—visit the Yasunkuni shrine honoring Japan's war dead, even though it included the graves of fourteen leading war criminals—a move an opposition politician likened to "dedicating flowers to Hitler."

As the April 24, 2001, date for the party primary approached, campaigning intensified. Minor candidates dropped by the wayside, and the contest narrowed to Hashimoto and Koizumi. Excitement spread through the LDP ranks as the Koizumi message hit home.

Hashimoto's minions found greater and greater resistance among the primary voters. One supporter told a campaign meeting, "Even if I asked them to vote for [Hashimoto], they don't give me a clear response." Others felt that their earlier optimism was starting to look unrealistic. Newspapers reported that "some dissident junior lawmakers in the Hashimoto faction, which had long boasted iron-clad solidarity, say they may not back Hashimoto and are calling on their colleagues in other factions to vote based on personal opinions rather than on factional lines."

Koizumi was on a roll. Party leader Shizuka Kamei noted that "he looked like a white knight on a ride for justice. That mesmerized our members."

Seeing the enthusiasm Koizumi had built among the grass roots, Hashimoto resorted to old-style patronage politics and sought support from other factions by offering top jobs to their leaders. Increasingly frantic to defeat Koizumi, Hashimoto's people reached far down into the special-interest/industrial groups looking for support. A Japanese newspaper reported that postal workers, whose 240,000 members formed the LDP's single biggest support group, were asked to return their blank ballots blank, so that Hashimoto's name could be entered on them by party workers.

Unfortunately for Hashimoto, Koizumi was able to amass a large lead in the rank-and-file voting. Party elders were still holding out hope that their rigged voting system would deliver the election to Hashimoto; after all, the new wild-card party members accounted for only 141 of the 487 votes that would determine the new party leader. But a new dynamic was at work. So unaccustomed were the leaders of the LDP to the limited party democracy that they overlooked the implications the rank-and-file vote would have for the rest of the process. Simply put, if the party members from a district overwhelmingly backed Koizumi, how could the Diet member who represented them (and depended on their support) ignore their wishes and back Hashimoto?

As the election neared, Diet members began to get the point. "I will vote for Mr. Hashimoto if my hometown LDP members choose him," one member of the Lower House Hashimoto faction said. "But if Mr. Koizumi comes in first, I may cast my ballot for him."

Blind to the end, a senior Hashimoto faction member wondered aloud whether Hashimoto's leadership would be "tainted by charges of being saved by factional interests if he trails Koizumi in the local votes but wins the overall race."

He needn't have worried.

Koizumi swept the rank-and-file vote in virtually every constituency. He won an astonishing 87 percent of the vote of the party members, guaranteeing him 123 of the 141 votes the members controlled. In only two districts, strongholds of the Hashimoto faction, did Koizumi fail to win the delegates.

While theoretically Koizumi could have been defeated by the 346 Diet members, the psychological impact of his incredible grassroots victory was so strong that no resistance was possible. Facing a general election at the end of July, three months after the election of a party leader, the LDP Diet members would not deny the popular Koizumi his place in the sun. As one LDP Diet member explained, "we picked the best man able to lead us into the election and that was it."

Even so established a figure as the chairman of Mitsubishi, Minoru Makihara, was swept up in the enthusiasm of Koizumi's victory. "The fact that the rank-and-file Liberal Democrats cast ballots overwhelmingly in favor of Koizumi is a clear signal to the Liberal Democratic Party that they have to change," he said. "It's a strong signal, and I'll hope they'll react."

In the end, the LDP Diet members ran for cover in light of

Koizumi's populist surge, and tamely ratified the vote of their grass roots. Koizumi got 175 of their votes, Hashimoto 137, and 34 went to for other candidates. Junichiro Koizumi was elected the twentieth president of the Liberal Democratic Party—and was anointed its choice for prime minister. It was the first time the LDP had chosen a leader who was opposed by the party's largest faction.

Koizumi overcame a stacked deck by scaring the daylights out of the party professionals with a landslide victory that came directly from the voice of the people. His story speaks to one of the most basic of all political principles: Those who would hold power must respect the overwhelming sentiment of their constituents—whether the constituency in question is a new political movement or an emerging commercial market.

Koizumi shook things up as soon as he took office. He began his administration with a sharp break from the LDP custom of awarding cabinet positions to each of the political factions, in proportion to their power in the Diet. It was a system that effectively delegated the choice of cabinet members to the factional leaders, making them more powerful than the prime minister in selecting his own cabinet. But an emboldened Koizumi promised to "destroy" forces standing against his reform agenda and to form a governing cabinet free of control by the party factions. "I will appoint the right people to the right posts," he said, "and do away with faction-oriented politics. My political life will end if I fail to form a cabinet that reflects the voice of the people. . . . I will never give in to any pressure. . . . Although the election will end . . . the real fight will not end until the voice of the people destroys forces that are resisting reforms."

Koizumi kept his word, ignoring the Hashimoto faction in favor of new cabinet members for ten of the seventeen posts, among them five women and three people with no political background at all. His most important wake-up call to the old regime was his choice of Makiko Tanaka, a popular but unconventional woman, as his foreign minister. Koizumi called his new cabinet "a national salvation," and to many it was.

As the new cabinet took office, Koizumi egged on his supporters with the cheer "Gambaro!" which translates, in the vernacular, as "let's close ranks and kick some butt." He was ebullient, remarking, "it seems the earth is shaking. The shimmering magma is about to explode." And he predicted that the LDP "will strike out its chest and move strongly toward the future."

Koizumi had accomplished his first goal—to seize control of his own party. Now he faced an even trickier hurdle: keeping the party united behind him as he moved toward reform.

To keep the party's Diet members with him, Koizumi concentrated his fire on the Hashimoto faction—the leading, but not the only, old-guard faction in the LDP. This group of elderly politicians had long manipulated the LDP through its unified determination to hold power. Counting almost a third of the LDP's Diet delegation among its members, the Hashimoto faction was the tail that had wagged the LDP dog for decades.

Yet Koizumi had clipped the old faction's power by defeating its leader, and now he mobilized his political constituency against the Hashimoto faction in the name of reform. The changes he advocated amounted to "an assault on the Hashimoto faction's financial interests," striking at the core of those who underwrote and funded the Hashimoto faction's political power.

His special target was the faction's relationship with the construction industry. Through its ability to produce huge annual budgets for public works—needed or not—the Hashimoto politicians had long guaranteed a rich and reliable flow of campaign contributions. Now, as prime minister, Koizumi pushed his agenda of cuts to the public works budget, limiting government debt, and diverting funds from the Roads Corporation (which used all car and gasoline tax revenues to build highways) to other parts of the government.

Hashimoto's people also depended on banks for much of their power. Here, Koizumi's aggressive plans to force banks to write off bad debt—which will likely force several major banks to close—struck at Hashimoto's vital power base.

As he cut off the oxygen supply to the Hashimoto faction, Koizumi was careful to romance the other LDP factions, whose Diet members had reluctantly voted for him in the party elections in deference to the will of their constituencies. Rewarding them with important cabinet positions, he took care to consult them on key decisions. But he also used realpolitik to compel their support.

Koizumi had his first major political test as party leader just three months after he took power, when he faced an Upper House election in which the LDP feared it might lose control to the reform-oriented DPJ. While the Upper House did not have much formal power, the LDP would lose tremendous prestige if it could not win the election in July 2001.

Now Koizumi used his enormous popular appeal to seal the loyalty of the LDP Diet members by throwing his weight behind *them* in the Upper House elections. Koizumi made television commercials for the LDP candidates and campaigned avidly for his party's ticket. It was an exceptional commitment—especially given that almost one-third of the LDP candidates were Hashimoto-faction candidates selected before Koizumi's ascent to power. Even as he was cutting off its funding and reducing its power, Koizumi held the Hashimoto faction hostage by making it clear that, without his support, their candidates for the Upper House would lose.

Koizumi had another ace in the hole—the Japanese parliamentary system. As prime minister he also had the power to decide when to schedule elections for the Lower House of the Diet—the one that really counted—whenever he wished. If Koizumi were to dissolve the Lower House and order new elections, he would be able to pick and choose which LDP members to support in the ensuing election. With his enormous popularity, this very prospect was enough to keep fearful LDP politicians in line. As Kaoru Okano, a former president of Meiji University, noted: "LDP members, especially the Lower House members, will not be able to openly criticize Koizumi for fear that they might not be authorized" to run as LDP candidates.

As Koizumi put it, "The LDP can carry out reform. That's why they made me the president. But if they ever try to stand in my way, there will be a Lower House election."

Armed with this political power, Koizumi was a confident man. "As long as I have more than 50 percent support from the people, I can [accomplish my reforms]. That there's opposition makes me feel it's all the more worth doing. . . . Everyone thinks that the [LPD] will drag me back or that opposing forces will resist to the end. [But] LDP legislators are surprisingly sensitive to public opinion. I think the LDP will have to follow me."

The other problem the LDP old guard faced was the same one that kept Blair's union leaders and leftists in line—they had no place else to go. The DPJ, the leading opposition party, was anathema to them. Its program of reform was even more radical than Koizumi's, and they could expect little mercy were they to try to leave the LDP and join the DPJ. Trapped in the LDP and terrified by Koizumi's popularity and power, the old guard stayed in the party and never ventured outside.

Having accomplished his first two objectives, then—taking power and keeping his old enemies within the fold—Koizumi was well-positioned to carry the day in the Upper House election.

In a speech designed to rally voters behind the LDP, the new prime minister promised: "I will carry out reforms that no other parties have dared to touch. . . . If the LDP wins this election, we will surely put them into action. . . . In the past, we could not propose something that caused pain. I dare to propose bearing pain now for a better future. I will enact bold and flexible reform, without fearing pain, without flinching at barriers of vested interest, and without being shackled by the experience of past. . . . There may be those [in my party] who oppose my reforms, but the majority support them. But unless people support the LDP, we cannot carry them out."

When the ballots were counted in the Upper House election, the LDP swept to its biggest victory in a decade, winning 65 of the 121 contested seats outright and 78 with its coalition partners.

Will Koizumi be able to pass his reforms, restructure Japan's economy, and maintain the loyalty of his captive party? Will he retain his popularity as his reforms cause higher unemployment? We don't know. But Koizumi's success in reforming his party, capturing its leadership, and uniting its disgruntled former leaders behind him contrasts sharply with the experience of Democratic Party leader George McGovern in the United States in 1972.

MCGOVERN REFORMS
HIS PARTY . . . AND
THE EMPIRE STRIKES BACK

All reformers do not go to heaven. Sometimes the one who succeeds in reforming his party is punished, not rewarded, for his efforts. Whereas Tony Blair and Junichiro Koizumi went on to lead their "new" parties to victory, George McGovern had a more lamentable experience. He sought to reform the Democratic Party, and succeeded beyond all expectations. He used the reforms to win his party's nomination for president. And then those whose power he had joyfully diluted got revenge by turning on him in the general election. The British labor unions stuck with Blair, as the Japanese bureaucrats did with Koizumi. But the American unions and the Democratic Party bosses thanked McGovern by stabbing him in the back—even though it meant reelecting Richard Nixon in 1972 for another term as Republican president.

Why did McGovern meet with such a fate? His reforms were overdue. They democratized the Democratic Party, and even shamed the Republicans into following their lead. The power of the oligarchs who had ruled politics for centuries was permanently curbed. But in their last gasp, in their death throes, they killed the reformer.

McGovern was right in judging that he would appeal to independents if he were able to emerge from his struggle with the party elders with the blood of the dragon on his sword. But he found out that in party politics, dragons don't really die. They get wounded, weakened, and angry. Then they come back. When the hero emerges to face the electorate in triumph, he must take care when turning his back on the dragon he has just defeated, lest the dragon start thirsting for a rematch. After all, in politics, the most lethal wounds are inflicted from behind.

How did McGovern go wrong where Blair and Koizumi got it right?

Ever since the days of Andrew Jackson and Martin Van Buren, political bosses have dominated American politics. Their collective motto, "to the victor belongs the spoils," rang like a clarion for seven generations of politicians as they lusted after the grand prize of the American presidency. In the 1930s, labor unions joined Democratic bosses as arbiters of their party's destiny, the judges who awarded the spoils of victory.

The system produced good presidents as well as bad ones. And the bosses who profited from it defended it zealously.

But the party-boss system had also produced the leadership that led America into a no-win war in Vietnam, and into a period of rancor and division that reached a crisis point with the election of 1968. As sentiment mounted against the mindless slaughter of American soldiers overseas, young political activists sought to use the democratic process in the Democratic Party to bring the troops home and end the war. But, to their collective discomfort, they soon found that the party was Democratic with a capital 'D' only. The only real power in its ranks belonged to the unions and the bosses, who were unshakingly committed to President Lyndon Johnson, to his designated successor Hubert H. Humphrey, and to the unpopular war.

The activists launched a grassroots campaign; abandoning beards, long hair, and drugs, they stayed "clean for Gene" as they went door to door campaigning for the antiwar senator, Eugene McCarthy—and scored upset victories in the primaries. When Senator Robert F. Kennedy later found it politically safe to enter the race, he jousted with McCarthy in the primaries, both pushing the agenda the activists had put on the table.

But there was one man missing from the primaries: the frontrunner, Vice President Hubert H. Humphrey. Abjuring the primaries for the very good reason that he would have gotten clobbered, Humphrey sat on the sidelines piling up endorsements from the bosses. The power of the old oligarchy that ruled the party was in full view, its fist no longer encased in velvet. The Humphrey delegates were all bought and paid for by the party and union bosses who had hand-picked them. As Theodore S. White observed, "Humphrey had entered no primaries . . . but the AFL/CIO . . . had delivered to him almost all of Pennsylvania, Maryland, Michigan and Ohio."

When Bobby Kennedy was assassinated on the night he won the California primary, the cause of the activists was stopped in its tracks. With Kennedy gone, the well-oiled party machine extinguished all dissent and steamrollered the Humphrey nomination. The old politics of the bosses confronted the new politics of the young activists and defeated them—not at the ballot box, but at the rigged Democratic National Convention in Chicago.

As the votes of millions in the Democratic primaries were overlooked and brushed aside by the bosses, the Chicago convention demonstrated that the Democratic party was a "politician's party." As White observed, "From clubhouse to city hall, from state capitols to Washington, the professional politicians all agreed: amateurs had to be kept out. Politicians controlled nominations, from local judges and aldermen to Senators and Presidents—and the only gateway to November elections lay through their anterooms."

But the young would not be deterred. Defeated within the Chicago convention hall, they literally took to the streets to manifest their anger at the proceedings inside. As the convention droned on to its preordained conclusion, young people who had worked heart and soul for McCarthy and Kennedy demonstrated in the streets. Chicago mayor Richard Daley's police responded with tear gas, fire hoses, clubs, and fists. What had begun as a largely peaceful, if spirited, demonstration dissolved into a police riot as idealistic young men and women shed their blood on Daley's streets and in Daley's parks at the hands of Daley's cops. Before television cameras, the young demonstrators shouted, in chorus, "The whole world is watching"—and indeed it was.

As the nation watched in shock at the bleeding, bandaged heads of its young in the streets of its second largest city, liberal senator Abraham Ribicoff of Connecticut—a former member of JFK's cabinet—

took to the convention podium to call for a halt to the police brutality. From his seat on the convention floor, Mayor Daley stood and flipped the finger at the senator on national television. Ribicoff had been scheduled to give a nominating speech for South Dakota senator George McGovern, a last-minute entry into the race. Throwing away his prepared speech, Ribicoff took Frank Mankiewicz's advice to talk "about what is going on in the streets." "With George McGovern," Ribicoff said, "we wouldn't have Gestapo tactics in the streets of Chicago." From the audience Ribicoff—and America—could see Mayor Daley screaming "Fuck you, Fuck you" repeatedly; "How hard it is to accept the truth," Ribicoff replied, "How hard it is."

McGovern recalled in his autobiography, "Abe and the mayor were shouting at each other and I knew the Democratic Party was being torn asunder." As Americans watched, Joseph Alioto of San Francisco nominated Hubert Humphrey. White captured the televised spectacle: "Back and forth the cameras swung from Alioto to pudgy, cigar-smoking politicians, to Daley, with his undershot, angry jaw, painting visually without words the nomination of [Humphrey] . . . as a puppet of the old machines."

Meanwhile, as Eugene McCarthy tore up the bedsheets in his hotel room to make bandages for the wounded demonstrators, Humphrey watched aghast as his chances of victory in November ran down the drain.

What was exposed in 1968, in short, was the authoritarian rule that lay at the core of the so-called democratic process. It not only kindled a public clamor for reform, but it made the rank and file squeamish to see their party, which had always stood for democracy, so nakedly exposed as authoritarian.

Humphrey went down to a well-earned defeat in the ugly wake of the 1968 campaign, and Richard Nixon was propelled to an ill-deserved victory. But the need for reform remained.

In truth, the system's flaws were evident enough to party insiders that a reform initiative had been launched before the Chicago convention had even opened. Insisting that the Democratic Party had become a closed society for party bosses, the reform forces brought a challenge before the convention's Rules and Credentials Committees, protesting procedures that permitted a handful of bosses to name candidates with no regard for the will of the voters in party primaries. Eugene McCarthy's representatives protested the bosses' control of the convention through the "unit rule," which allowed a majority of a state delegation to muzzle the minority and require it to vote the way the majority instructed. Meanwhile, members of the Credentials Committee found it obnoxious that some delegates to the convention had been selected two years earlier, before the candidates were even known, and that in some states one man had the authority to select delegates. Many delegates were chosen by arcane or "incomprehensible" rules, the committee found; some were chosen in private homes, and in many cases race was used as a criterion for exclusion.

Before the Chicago convention Governor Harold Hughes of Iowa recommended that the party establish a commission to review the delegate selection process and reform it in time for the 1972 convention.

But the Credentials Committee, firmly under Humphrey's control, turned down the reformers' proposals. To have accepted them would have endangered their ability to deliver the nomination to Humphrey.

But, in a concession to the reformers, some of the Humphrey delegates were willing to consider rules changes—as long as they took effect at the *next* convention. On the floor, the delegates narrowly passed a historic resolution—by 1,305–1,206—to require reforms at subsequent party conclaves.

The resolution required that "all Democratic voters have had full and timely opportunity to participate" in choosing future delegates, and in naming the candidates for whom they would be pledged. It required that "all feasible efforts [must be] made to assure that delegates are selected through . . . procedures open to public participation within the calendar year of the national convention." The resolution assured that, even after the disaster of Chicago, the process of reform would go on.

Democratic National Committee (DNC) chairman Fred Harris, a reform-minded senator from Oklahoma, duly named a Commission on Party Structure and Delegate Selection. When he appointed Senator George McGovern of South Dakota to be its chairman, it was clear he meant business.

Always a liberal and an outsider, McGovern had been one of the few who dared to speak up against the communists witch-hunts of the 1950s. McGovern dissented, charging that "anyone capable of recognizing the devil" would know it was the investigators, not the investigated, who posed "the real threat to the United States."

McGovern felt that American foreign policy during the Truman years "was needlessly exacerbating tensions with the Soviet Union," and admitted he "wasn't happy with the direction the Democratic Party was taking." In particular, he rejected the influence of organized labor. In the 1960s the South Dakotan McGovern alienated the unions by attacking their refusal to load American wheat on ships bound for the Soviet Union, then in the grip of famine. Anxious to relieve hunger abroad and enrich his farmer constituents at home, McGovern also opposed union-backed legislation that would require half of U.S. grain exports to be shipped in American boats. Labor leaders felt they had been "double-crossed" by McGovern. Chief among them was George Meany, the head of the AFL-CIO, whose dislike of McGovern bordered on the pathological. When McGovern came out against the war in Vietnam, Meany practically accused him of treason.

McGovern had declared his candidacy for president after Bobby Kennedy's murder, in order to give Kennedy delegates an alternative to McCarthy or Humphrey. The two-month campaign he waged had little hope of capturing the nomination, but it managed to solidify McGovern's position on the party's left. Now, his selection as chairman of the Reform Commission was intended to demonstrate the seriousness of its purpose and the sincerity of the party's intention to reform.

As he assumed the leadership of the commission, McGovern initially sounded a pragmatic note, emphasizing the need for reform to save the party. At its first meeting, on March 1, 1969, McGovern said, "When parties have been given the choice of reform or death in the past, they have always chosen death. We are going to be the first to live." Championing "participatory democracy . . . as a strategy for unifying the party," he contended that "the party would only unite behind a nominee selected democratically by rank-and-file Democrats," and that "never again would the Democrats accept a candidate hand picked by oligarchs of the party."

The commission McGovern chaired was clearly liberal, in its

members and its goals: it included a leader of the Mississippi NAACP, a liberal former governor of Florida, Indiana's progressive senator Birch Bayh, and other visible Democratic liberals. But it was seeking to reform a Democratic Party whose hubris was still intact. Having won seven of the past ten national elections, and controlling the White House for twenty-eight of the preceding thirty-six years, the party didn't see the need for reform.

Steelworkers Union leader I. W. Abel spoke for the bosses when he declared that "the party which nominated Roosevelt and Truman does not need reforming." Abel, appointed to the Reform Commission as a token representative of labor, boycotted its proceedings.

McGovern himself thought the chairmanship "a thankless task." His commission of liberals, appointed by one liberal leader and chaired by another, was setting its sights on an ambitious course of reform, without the support of the mainstream of the party. All that McGovern could see "getting out of it was irritating a number of people." And he was right.

As soon as the commission moved into its offices in Washington's Watergate Hotel, "the flak began to fly" from organized labor. George Meany was reportedly " 'livid' over the very notion that the party was in need of reform" and "the unions proceeded to boycott the commission's hearings."

Hostility toward McGovern's commission spread quickly. Georgia's racist governor, Lester Maddox, mixed up his body part metaphors, calling it "an arm of the socialist wing of the Democratic Party." Meany also opposed the commission's efforts to increase representation of women, youth, and minorities among the delegates. As unpopular as reform was among party leaders, McGovern made it more so by alienating them gratuitously. Almost proud of the enemies he made, McGovern declared that "we have to be willing to make somebody mad, and to take some political risks, if we are to reclaim our society."

When the beast himself, Mayor Daley, attended the commission hearing in Chicago in mid-1969, he surprised many by testifying in favor of reform and backing delegate selection by state primaries. But rather than pocket this significant concession, McGovern couldn't restrain himself from pushing Daley, demanding that the mayor drop criminal charges against the 1968 Chicago protestors. Prodded beyond endurance, Daley exploded, vowing: "if you're asking for amnesty for violations of the law, I'll never be a part of it." The headlines in Chicago the next day reported that Daley had "rebuffed the reformers."

McGovern later regretted his mistake; in his autobiography, he admitted that "it would have been better simply to commend the mayor for endorsing our reform efforts and leave the matter of how to handle the Chicago disturbances to local authorities." But as the commission pursued its schedule of hearings and deliberations, this kind of faux pas did nothing to endear the cause of reform, already unpopular with the regulars, to the broad center of the party. McGovern seemed oblivious to even the most elementary protocol. Party chieftains claimed that he failed to notify them about meetings in their own backyards; Texas governor Preston Smith learned about hearings in his own state by reading the newspaper. Democratic National Committeeman J. Marshall Brown of Louisiana "was so infuriated at the commission's plans that he 'ordered' McGovern to stay out of his state."

Still, McGovern didn't get it. While he admitted that "the commission offended some leaders," he also argued that "the party's shortcomings are too important to ignore." Yet McGovern's own shortcomings were, on balance, equally troublesome and hard to ignore. The comparison with Tony Blair is instructive: Blair wore out union-office welcome mats with his incessant consultations with the power brokers he was seeking to dethrone. These cannot have been pleasant; Blair must have felt like Daniel wading into the lions' den as he faced the inflexible leftists and union leaders. But he made a sincere and concerted effort to bridge his differences with his rivals. When they disagreed with him, his opponents at least understood what he believed, and why.

McGovern, on the other hand, did little to court the leaders of his own party and convince them to join him. His contacts with them were stilted, artificial, and infrequent. At times he seemed convinced that he would catch an infectious disease if he spent too much time with the power brokers. He refused to hold the tough meetings, and was content to treat party leaders as arm's-length adversaries. His snobbish attitude was evident in an account he gave of his meeting with Mayor Daley's Democratic committeemen. McGovern remembered "noticing how much they all looked alike as they applauded in tandem. They all seemed to be middle-aged, red-faced, heavy-set males. There was a lot of cigar smoke in that room, a lot of beef." His contempt for established white, male party leaders must have been unmistakably obvious; McGovern's conduct ensured that even if his reforms should succeed, he never would.

So many men and women in politics, business, academia, or civic life have no stomach for facing controversy directly and personally. They value the antiseptic warfare of politics at long range, and shy away from face-to-face confrontation or even negotiation. But the experience of a thwarted reformer like McGovern proves that direct dialogue with those in power is essential. Even as he pursues his agenda, the successful reformer must spend time persuading opponents and reconciling his adversaries to his actions if he wants to preserve his own future. In the delicate process of reform—regardless of the venue—everything depends on diplomacy.

When McGovern's commission finally made its recommendations, they were every bit as earthshaking as he had hoped; in the end, they would change the face of American politics. The commission began with a catalogue of abuses that would stun any modern observer. The old rules were so arcane and biased toward the white, male party establishment that some states operated without any written rules; others excluded voters from any role. California had a winner-takes-all primary with no provisions for the losing delegates to participate. All in all, the rules were simply intended to maintain the status quo and keep new ideas, new people, and new perspectives out of the process. The way in which we choose party nominees still has its flaws: Money has too much power, momentum counts for more than it should, and there is not enough time to test each candidate. But at least the nominations are democratic and the voters not the bosses control the outcome. To read the findings of the McGovern Commission is to look into another era in our politics, only thirty years past.

But no abuse was as important as the unit rule, which provided for majority domination at every level of the process. If a local town Democratic committee voted 51–49 for a candidate, all of its votes went to the winner. By applying the unit rule at each level of the delegate selection process, even widely popular "minority" views could be completely deprived of any representation at all.

The commission's reforms were radical and sweeping, outlawing secret ballots and caucuses. Delegates were not to be selected before the year of the convention. The unit rule was abolished, and all delegates would henceforth be chosen by some electoral process, be it a primary or a party caucus open to all members. State parties would be obliged to "overcome the effects of past discrimination by affirmative steps" to encourage representation by minority, young, and female delegates.

Most of these reforms were grudgingly accepted by the party elite, but a further measure adopted by the commission stuck in its collective craw. Although senators, congressmen, governors, and high-ranking party officials had traditionally been selected ex officio as party delegates, the McGovern commission banned the practice. In the new system, you either won your seat in a primary or a caucus, or you watched from the gallery. The move was a direct slap at the party's top leaders, who felt, not surprisingly, that their elected positions should entitle them to seats at their own party convention.

Ultimately, it was this systematic exclusion of many of the Democratic Party's elected officials from the 1972 national convention that spelled doom for George McGovern. This blunder, which might have been averted had he developed a personal relationship with his own party's leaders, is the only one of McGovern's reforms that has subsequently been repealed. This gratuitous rejection of the party's elite would cripple his candidacy in 1972.

And there was one more reason for enmity as McGovern's findings became public: The new rules adopted by the commission were not just proposals—they were directives. In February 1970, the commission sent letters to Democratic state chairmen telling them how to comply, and the DNC included the guidelines in its preliminary call for the 1972 convention. Anyone who failed to follow the new rules wouldn't be seated as a delegate.

But where did this twenty-eight-member commission, stacked with liberals, get the power to dictate rules to the entire Democratic Party? There had been no party-wide vote or even a conference of delegates to ratify their proposals. Their power stemmed only from the reform resolution passed narrowly in the midst of the chaotic 1968 national party convention. This resolution, adopted as a sop to the reformers as an angry mob protested and battled police outside the hall, offered too thin a cloak of legitimacy for what was, in effect, a coup d'état in the party. At the time, most leaders saw the convention resolution as a palliative to the party's liberals—to ensure their loyalty after their hard-won primary victories were swept aside by the bosses. To use this slender reed to support a massive edifice of party reform was asking for trouble. For a commission bent on introducing a new standard of broad representation into its procedures, this was a strangely autocratic gambit—a far cry from the referendum Tony Blair used to achieve reform. The McGovern Commission's reforms may have been righteous, but they lacked the moral authority that only ratification by the voters could grant.

McGovern, nevertheless, was rhapsodic about the reforms. He used them—and his opposition to the Vietnam War—as the centerpieces of his campaign in 1972. "The Commission on Democratic Party Reform, which I chaired," he told a New York college audience, "has written new rules . . . which guarantee the most open and the most responsive convention. . . . In 1972 you will be heard." Addressing the Maryland State Legislature, he vowed that "we will not be weak-kneed on the hard question of political reform. . . . I do not ever again want to see another convention like the one in 1968." In his announcement speech, he told listeners, "We will not be helped . . . by leadership built on image-making or television commercials; [or] by those who seek power by back-room deals."

Had McGovern made a few back-room deals himself, he might have fared better in the election.

McGovern's first major adversary for the 1972 presidential nomination was Edmund Muskie, Humphrey's vice presidential running mate in 1968. But Muskie soon withdrew, leaving Humphrey himself as McGovern's opponent. As they faced each other, it seemed as if Humphrey, much like Muskie before him, was waging a completely different kind of campaign against McGovern. McGovern used his own commission's new rules "to the hilt," as he battled in the primaries, taking particular advantage of the requirement that all delegations had to include minorities and young people. Muskie—and then Humphrey—meanwhile, "continued to rely on the professionals" for support. But the pros could no longer deliver. The McGovern rules had changed the game. Bosses didn't matter much anymore; grassroots activists controlled the nominating process.

To add to the professionals' angst, the bosses themselves couldn't even get seats at their own party's convention. Due to the ban on automatic delegate slots for elected officials, hundreds of congressmen, senators, governors, labor leaders, and state party chairmen were forced to watch the 1972 convention on television. Their outrage boiled over when Mayor Daley and his Illinois delegation were tossed out of the convention hall after losing a credentials fight to the McGovern forces. To throw Daley out one day and then ask his help in carrying Illinois the next was a strategy not likely to succeed.

Angered, frustrated, spurned, and shaken, the party bosses vowed revenge against McGovern. As the South Dakota senator took the stage at 3:00 A.M. to address the exhausted throng at the 1972 Democratic National Convention—the battles over reform, credentials, party platform, and the nomination itself had lasted long into the night—he faced a party divided as it had not been since 1948.

His speech, uplifting to read, did little to assuage the embittered party regulars. Instead of healing divisions, McGovern used the occasion to further flay the very bosses whose support he would need to win the election. "In a democratic nation, no one likes to say that his inspiration came from secret arrangements behind closed doors," he chided. "The destiny of America is safer in the hands of the people than in the conference rooms of any elite." It's hard to imagine which must have been more insulting—McGovern's swipes at backroom machine politicians, or the fact that most of those same politicians weren't even welcome at their own party's convention to hear McGovern's sermon firsthand.

Even as McGovern was accepting the nomination of their party, the bosses, cursing under their breath, resolved to do everything they could to torpedo his campaign. And they didn't have to wait long. Just out of the starting gate, McGovern stumbled. It emerged that his running mate, Senator Thomas Eagleton of Missouri, had been hospitalized three times for depression in the previous twelve years and had been treated twice with electric shock therapy. Eagleton hadn't told McGovern, and when the newspapers broke the story, the candidate was blind-sided.

Nevertheless, McGovern loyally, if precipitously, jumped into the breach and announced that he stood behind Eagleton "one thousand percent." Though he conceded he hadn't known about the hospitalizations, McGovern declared "I wouldn't have hesitated one minute [in choosing Eagleton] if I had known everything Senator Eagleton told you today."

But the doubts about Eagleton grew, and when McGovern consulted his running mate's doctors, his concerns were not allayed. Asked how their former patient would perform if he had to become president, one of Eagleton's doctors said: "I don't like to think about that prospect." Another allowed that it "would make me most uncomfortable."

Alarmed, the establishment began to line up against Eagleton. The *New York Times* called for Eagleton to leave the ticket. "It is painfully evident," the editorial board opined, "that . . . little attention was given by Senator McGovern to learning all he should have known about his running-mate's strengths and weaknesses."

Soon afterward—just eighteen days after he had been nominated—Eagleton was forced off the ticket, the first time in history that a vice presidential nominee had withdrawn. McGovern chose Kennedy heir Sargent Shriver to replace Eagleton, but he later admitted that "the . . . matter ended whatever chance there was to defeat Richard Nixon in 1972." Eagleton, for his part, said the episode was only "one rock in the landslide." McGovern countered: "Perhaps that is true but landslides begin with a single rock."

Some were appalled at McGovern's lack of wisdom in choosing Eagleton. Others were revolted by his lack of loyalty in dumping him. Many were angered by both. The Missouri State Democratic chairman condemned McGovern's "very, very shabby treatment" of Eagleton, and added, "this fall's election will probably find me and my friends spending our money and working to elect our state, congressional and local candidates" rather than throwing their efforts behind McGovern.

Like a bleeding bather, McGovern now attracted the political sharks who had been out for his blood from the start. The labor leaders and Democratic bosses got their revenge. While McGovern, oblivious to their animosity, said he wasn't too concerned about getting the labor leaders' support, the head of one major union said that labor "wouldn't touch [McGovern] . . . with a fifty-foot pole."

McGovern's chief nemesis, as expected, was George Meany, normally the chief booster for Democratic nominees. Calling McGovern a traitor because "he don't stick to his people," Meany even extended an olive branch to Nixon. "I want you to know," he told the Republican president, that "I am not going to vote for McGovern or for you. But you'll be doing all right with the Meany family," since, he said, his wife and two of his daughters would vote for Nixon. For the first time in seventeen years, the AFL-CIO would not endorse the Democratic candidate.

Since only three of the thirty Democratic governors had endorsed McGovern before the convention, the candidate might have been expected to work overtime to win their loyalty. But he didn't; that wasn't his style. Chief among McGovern's critics was the future president, Georgia's governor Jimmy Carter.

Ambitious to make a name for himself in the center of the party, Carter helped lead an "ABM" movement—Anybody But McGovern. Carter, speaking "for the large block of Southern Governors," said that McGovern's views were "completely unacceptable to the majority of voters in many of our states."

Among the group of unenthusiastic Democrats was former president Lyndon B. Johnson. In a letter to party leaders, he confirmed that he would support the ticket because he had been a Democrat for so long. But he pointedly failed to back McGovern by name, and even went so far as to remind Democratic voters that how they vote "in a presidential campaign is a matter of conscience."

Nixon later wrote that he was grateful for the letter—as he should have been.

McGovern made a personal pilgrimage to Johnson to ask for his support, but the former president avoided making a campaign appearance for the man who had been among his most persistent critics. McGovern even made a last-minute appeal to Mayor Daley, trying to "patch up matters with the traditional sources of power whom the new politics people had offended." Daley gave McGovern a pro forma endorsement, but otherwise sat on his hands throughout the fall.

As election day neared, it seemed the only person in Washington who was happy with McGovern's behavior was Richard Nixon. Because McGovern had alienated key Democrats, they were increasingly receptive to the incumbent's reelection campaign. The former governor of Texas, John Connally (a Democrat who had been wounded when President Kennedy was killed) endorsed the Republican president, and headed a Democrats for Nixon Committee. Plenty of other Democratic leaders harbored similar feelings: Moderate and sometimes even conservative in their ideology, they didn't really mind seeing Richard Nixon in the White House. In fact, most of them were more comfortable with him than they ever would have been with McGovern. At least he would play ball with them—even if he was on the other team.

On election day, McGovern was massacred. Nixon got forty-seven million votes to his twenty-nine million. McGovern won only liberal Massachusetts and black Washington, D.C. He didn't even carry his home state of South Dakota.

McGovern's story, ultimately, has some of the overtones of the classic tragedy. His reforms were wise and badly needed; because of them, the candidate selection process at all levels of American politics is basically democratic. Even the Republican Party, not to be outdone by its rival, adopted these reforms in McGovern's wake. Yet McGovern himself was felled in the process—not because of the reforms themselves, but because his own hubris ultimately exceeded even that of the party bosses he sought to unseat. In a critical moment in his party's history, George McGovern proved a divider, not a uniter. And in the end he dragged his party down with him to defeat.

STRATEGY FIVE

USE A NEW TECHNOLOGY

Marshall McLuhan, the philosopher and media guru, hit upon an undeniable political reality when he memorably declared, "the medium is the message."

Each new form of communication, in politics as in the rest of life, has an essential character. The politician, the businessman, the leader who grasps its potential can use its power to sweep all before him. As each technological innovation opens up new means of communication, it offers opportunities to convince, to persuade, and to move the people in one direction or another.

Every decade or two—since at least the turn of the twentieth century—a new communications tool has emerged and revolutionized the relationship between voters and candidates. And the politician who was astute and adventurous enough to seize it has usually been able to ride it into power.

But the key is not just to use the medium, but to understand it—to grasp its "message," and the inescapable impression the media conveys, almost regardless of what is actually being said. A speech made on a public platform before a large audience, for example, conveys an undeniable message of prestige and authority. Every medium has a comparable inherent message.

Some of the most effective communicators of the modern era

have succeeded in understanding the unique qualities of a new medium—and marshaling its power to serve their goals.

Franklin Delano Roosevelt used radio as his medium and intimacy as his message. Possessed of a rich and resonant voice, FDR was the first leader to give the presidency a sound to go with the images voters had grown accustomed to seeing in newspapers and silent newsreels. Using the growing phenomenon of national radio networks to broadcast both his public speeches and the more personal addresses he called "fireside chats," he came into the living rooms of every American, seeming to speak to each listener individually with a blend of casual candor and inspiring eloquence.

Opening his intimate chats with the words "My friends," he reached out to Americans in a way no president had ever done—and built a bond with the voters that lasted through four successful presidential races. By moving political discourse off the platform and into the living room, Roosevelt was able to spark a level of connection with the average person more intimate than any other politician ever had.

Both John F. Kennedy and his rival Richard Nixon used the medium of television heavily in their 1960 campaigns for the presidency. Nixon had actually been a kind of pioneer in the use of television, using it to save his political neck in 1952 in a televised plea to the nation—called the "Checkers speech"—to overlook a financial scandal that threatened his career. But Nixon tended to treat television as radio with pictures—a vehicle for delivering a textual message. It was Kennedy who realized that the image was everything in the new medium. By offering glamour where Nixon merely gave speeches, JFK forged a charismatic relationship, particularly with America's young, that remained fixed in their hearts and minds long after Kennedy was dead. Could JFK have ever been president without the tube? Not likely. They were made for each other.

Kennedy's successor, Lyndon B. Johnson, had a different kind of problem. A compelling presence in person, Johnson—a big, ungainly, coarse-mannered Texan—could not hope to capture hearts and minds the way Kennedy had. So Johnson invented a new medium to promote his candidacy—the thirty-second negative TV ad. Unable to press the flesh of every voter in the United States, he found another way to reach them: by stoking the flames of fear inspired by his rival's extremist rhetoric. Whereas other politicians used television to sell themselves or their positions, Johnson used it for something quite different—to inspire instant emotional reactions from tens of millions of voters.

The common denominator among these new technologies and techniques is that they revolutionized politics by bringing a secret weapon onto the field, which only one side knew how to use effectively. Like the first warriors who met bows and arrows with gunfire—or like America in 1945, using the atomic bomb to bring an end to the war with Japan—FDR, JFK, and LBJ each stole a march on their rivals by harnessing a new medium and putting it to use at a time when their political rivals did not understand how to use it.

Actually, American history is filled with the stories of leaders who used new forms of communication to dominate or change the political process:

- Benjamin Franklin used his ownership of a printing press, and his publication of an almanac and a daily newspaper, to bring his political views to a wide new audience.
- Thomas Paine exploited the new strategy of pamphleteering to broadcast the reasons for American independence in his tract *Common Sense.*
- Andrew Jackson and Martin Van Buren invented the machinery of the modern political party and used it to win the election of 1828.
- The Whigs countered by using mass parades, festivals, fairs, and hoopla on a grand scale to recapture the presidency in 1840.
- Abraham Lincoln used open letters to newspapers and speeches to the public to frame his case for the Union in a way earlier presidents had never done—or had to do—before.
- Theodore Roosevelt's energetic regimen of public speaking—his virtual invention of the presidency's "bully pulpit"—transformed the office after decades of elusive and passive presidents.
- Woodrow Wilson broke down a symbolic barrier between two branches of government by becoming the first president to address Congress in person.
- Harry Truman's "whistle stop" campaigning brought personal presidential barnstorming to a new and frenzied pitch.

- Ronald Reagan's merger of theater and politics, bringing the skill of an accomplished performer to the presidency, earned him a justified reputation as "the Great Communicator."
- The permanent campaign of Bill Clinton, using polling and television advertising during his term to project and explain his issues, rescued his faltering presidency again and again.

As we ponder modern politics, it is becoming increasingly evident that the Internet will offer yet another opportunity for a great leap in communications, bringing a customized one-on-one dialogue into the home of each voter. Instead of the one-size-fits-all television campaigns that dominate our politics, the Internet in general—and e-mail in particular—will make possible a conversation between each voter and each candidate about issues and ideas. The interactivity of the Web means an end to "I talk, you listen" politics, and the beginning of a two-way discussion as the basis of political communication. Some feel this is long overdue.

FDR USES RADIO TO
REACH AMERICA

The key in using a new technology involves taking the time to grasp the unique potential that it—and it alone—affords. Anyone can recognize a new tool for communication and exploit it for political success. It takes a truly skillful—and fortunate—politician to recognize the special qualities of a new medium, and use them effectively.

The phenomenon of radio had begun sweeping the nation in the 1920s. The fact that Franklin Delano Roosevelt used the radio to talk to voters was hardly surprising; anyone who was president in the 1930s would certainly have done the same. His brilliance was in seizing upon the medium's extraordinary potential for intimacy.

Voters accustomed to seeing their political figures speak from distant platforms before large audiences thrilled to the experience of settling back in their own living room to hear the president speak, as if he had dropped by for an evening visit. He seemed to invite people to lean back, kick off their shoes, and listen to him as he took the opportunity to explain complex issues in simple and easily understood terms. A nation eager for guidance and comfort found it in abundance in his warm, resonant voice and everlasting optimism.

Roosevelt realized that the emotional stress of the Great Depression, with the disaster it brought with it into each home, had caused people to yearn for an emotional connection with each other, and with their leaders. In an astounding piece of theater, this son of wealth and luxury made himself a welcome and trusted presence to millions of people who could just barely afford a radio. FDR made himself welcome not just as a politician, but as a friend.

Roosevelt became the first American politician to use this intimacy with the voters. More than three hundred times during his twelve years as president, FDR took to the microphone to speak to the listeners he greeted as "my friends." Whenever his administration faced a crucial juncture—first in the battle with the Depression, then in World War II—FDR's charismatic voice would come into homes across the land to explain to people what was at stake.

Other presidents had used radio before Roosevelt, but none had showed any skill or creativity in using it as a political organizing tool. And none had the dramatic presence of FDR. Wilson, Coolidge, and Hoover had all experimented with reaching the voters over radio, but none was able to establish the emotional connection with the American people that came so easily to Roosevelt.

Roosevelt did more than use radio to get elected and reelected (and reelected and reelected). He used it as an essential tool of governance. He knew that to fight the Great Depression that was choking America as he took office in the 1930s, he had to thaw the fear that had frozen commerce. No speech printed in a newspaper or delivered from a podium could do that. But, he realized, radio could.

By figuratively sitting by the fireside and holding the nation's hand, he could induce its citizens, investors, and consumers to regain their nerve and verve. It wasn't just that he could speak to them over radio: he could melt their fear with his warm, mellow voice, right in their own cozy sitting rooms.

As he sought to enact his legislative program, he enlisted the support of the American people by explaining the details over the radio.

With unemployment high, "countless citizens found themselves with hours of involuntary leisure time and a need for escape," media historian Robert Brown notes. Radio offered a reliable and constant source of culture, entertainment, and diversion.

Franklin Roosevelt was quick to perceive its importance. As governor of New York, FDR took full advantage of radio's potential, addressing the state more than seventy-five times in four years. He had an almost evangelical passion for the medium: Early on, FDR told NBC president Merlin Aylesworth that "nothing since the creation of the newspaper has had so profound an effect on our civilization as radio."

FDR grasped immediately that radio could bridge the growing psychological alienation between the governing class and the governed. Even more impressively, he understood the potential it afforded for intimacy. "Amid many developments of civilization which lead away from direct government by the people," he said in 1933, "radio is one which tends on the other hand to restore contacts between the masses and their chosen leaders."

In Roosevelt's fight to pass legislation to help New York deal with the Depression, the governor would use the radio to bypass Republican opposition and take his case directly to the people. "During each [legislative] session," speechwriter Sam Rosenman remembered, "he delivered radio talks in which he . . . appealed to [the people] for help in his fight with the Legislature. . . . A flood of letters would deluge the members of the Legislature after each talk, and they were the best weapon Roosevelt had in his struggles for legislation."

Roosevelt himself noted: "Time after time in meeting . . . opposition, I have taken the issue directly to the voters by radio, and invariably, I have met a most heartening response."

Through radio, Roosevelt developed his own citizen lobby. Stirred by his passion and heartened by his informative reports to them on the radio, the people of New York rallied behind him and lobbied his opponents in the legislature. Radio permitted this dialogue with the electorate, which had previously been impossible.

One of the reasons President Roosevelt loved radio is that most newspapers regularly opposed him, endorsing his GOP rivals and attacking his politics. Eager to go over their heads, he found in radio the perfect instrument to make his case directly to the voters. And this master politician took care to assure his popularity with the barons of radio. As European governments began nationalizing radio, Roosevelt assured American station owners that he would not follow suit. In return, just as FDR was taking office in March 1933, National Association of Broadcasters president Alfred McCosker pledged "radio's full and unqualified cooperation in the tasks before it." Throughout his presidency, Roosevelt would get radio time whenever he wanted it. (FDR also could use the stick to promote cooperation from radio executives. Any station that was not friendly "unexpectedly found itself devoid of a broadcasting license the next time it came up for renewal by the FCC," Brown notes.)

And there was another reason Roosevelt and radio were such a good match: his physical condition. Afflicted with polio in 1921, Roosevelt lived in a wheelchair for the rest of his life. Unable to walk, radio was the only practical way to project his presence throughout the country. It provided this disabled president with "the authority of his voice," he himself said, to use in leading the nation.

As the *New York Times* reported on March 13, 1933, "Radio has given a new meaning to the old phrase about a public man 'going to the country.' When President Wilson undertook to do it in 1919, it meant wearisome travel and dozens of speeches. . . . Now President Roosevelt can sit at ease in his own study and be sure of a multitude of hearers beyond the dreams of the old-style campaigners."

From the start, Roosevelt demonstrated that his would be a presidency defined by radio. Upon winning the Democratic Party nomination in 1932, FDR broke with precedent and flew to the Chicago convention to accept it in person. "The traditional practice had been for a nominated candidate to delay his acceptance speech until after having been personally notified by party delegates one to two months later." But FDR would make his own tradition. At a time when air travel itself was a novelty, speechwriter Sam Rosenman recalls that Roosevelt grasped the "value of drama in public office and in public relations." More to the point, he understood that the "dismayed, disheartened and bewildered nation would welcome something new, something startling, something to give it hope that there would be an end to stolid inaction. He wanted people to know that his approach was going to be bold and daring; that if elected he would be ready to act—and act fast."

It was in this acceptance speech, carried by radio throughout the nation, that FDR first used the phrase which was to define his presidency when he declared: "I pledge myself to a New Deal for the American people." A new political era was born.

In September of 1932, he made four national radio campaign speeches. And, on October 31, 1932—in what the Los Angeles *Evening Herald* excitedly called "the greatest debate in American political history," he and his opponent, Herbert Hoover, gave back-to-back national radio addresses. Hoover had the advantage of incumbency, but his uncommunicative glumness was no match for Roosevelt's mastery of the medium. Game, set, match to FDR.

After Roosevelt won handily, by 472 electoral votes to 59, Hoover grumpily blamed the new radio tool: "Roosevelt's method was to pound incessantly into the ears of millions of radio listeners, by direct statement and innuendo, the total heartlessness of his opponent."

It was a stunning victory—but far greater battles lay ahead. On taking office, Roosevelt faced the most abject crisis to confront any incoming president since Abraham Lincoln. Over the past three years, the nation's economy had shrunk in half, falling from $100 billion to a mere $55 billion. Unemployment had risen from 4 percent to 25 percent. One American in four was out of work.

The economic debacle demanded a leader. Past presidents had assumed relatively little responsibility for the nation's economy; to them, recessions, depressions, boom times, and the other ebbs and flows of the business cycle were the political equivalent of the weather—conditions that may affect an administration, but which the president could do little about. Yet after the Great Crash of 1929, Herbert Hoover's ineffectual passivity infuriated the American people as they suffered the bludgeoning of the worst years of the Depression. FDR knew instinctively that he would have to play a far more active role—to take the country's spirit in hand as no president before him ever had.

Radio, thus, was not just a political instrument for Roosevelt, but an economic stimulant—a tool to help overcome the lack of confidence that was at the root of the Depression. If his voice could rekindle faith and engender a feeling of hope and optimism, he believed, he might just be able to talk America back to good times. If FDR the politician realized that radio would help him to win the election, Roosevelt the economist grasped that it would also help to banish the national fear that held the economy down.

In his inaugural address, FDR told the national radio audience that "the only thing we have to fear is fear itself—nameless, unreasoning, unjustified terror which paralyzes needed efforts to convert retreat into advance." Radiating a confidence America had not heard for years, FDR predicted that "this nation will endure as it had endured, will revive and will prosper . . ."

The *New York Times* gushed: "the human voice . . . does reflect man's mood, temper, personality, and . . . character. President Roosevelt's voice reveals sincerity, goodwill, kindness, determination, conviction, strength, courage, and abounding happiness."

As he spoke these confident words, America's banking system was collapsing around his ears. From 1929 through 1932, five thousand banks had failed. With no program of deposit insurance, nine million savings accounts—totaling $2.5 billion—had gone down the drain. With the crisis worsening hour by hour, Roosevelt closed all banks in the nation for four days. In less than twenty-four hours, he got Congress to pass the Emergency Banking Act, permitting the reorganization and eventual reopening of banks across the country.

Turning to radio one week after taking office, Roosevelt explained his banking policies one week into his term, in the first of what would become his celebrated "Fireside Chats," on March 12, 1933:

I do not promise you that every bank will be reopened or that individual losses will not be suffered, but there will be no losses that possibly could be avoided; and there would have been more and greater losses had we continued to drift. I can even promise you salvation for some at least of the sorely pressed banks. We shall be engaged not merely in reopening sound banks but in the creation of sound banks through reorganization.

FDR used radio to bolster confidence in the nation's banks. "After all," he said,

there is an element in the readjustment of our financial system more important than currency, more important than gold, and that is the confidence of the people. Confidence and courage are the essentials of success in carrying out our plan. You people must have faith; you must not be stampeded by rumors or guesses. Let us unite in banishing fear. We have provided the machinery to restore our financial system; it is up to you to support and make it work.

The address was an amazing success. When banks reopened there was no panic, and the system survived. The historian David Kennedy relates in his book *Freedom from Fear* how "In a voice at once commanding and avuncular, masterful yet intimate, he soothed the nervous nation. His Groton-Harvard accent might have been taken as snobbish or condescending, but it conveyed instead that same sense of optimism and calm reassurance that suffused his most intimate personal conservations.

"On Monday the thirteenth the banks reopened, and the results of Roosevelt's magic were immediately apparent. Deposits and gold began to flow back into the banking system. The prolonged banking crisis . . . was at last over. And Roosevelt . . . was a hero." Even inveterate Roosevelt-hater William Randolph Hearst was moved to tell him: "I guess at your next election we will make it unanimous."

The commentator Edwin C. Hill wrote, "It was as if a wise and kindly father had sat down to talk sympathetically and patiently and affectionately with his worried and anxious children, and had given them straightforward things that they had to do to help him along as the father of the family. The speech of the President's over the air humanized radio in a great governmental, national sense as it had never been humanized before."

The political power of FDR's use of radio was apparent to all. The *New York Times* wrote: "His use of this new instrument of political discussion is a plain hint to Congress of a recourse which the President may employ if it proves necessary to rally support for legislation which he asks and which the lawmakers might be reluctant to give him."

His Fireside Chats were typically delivered at 10:00 P.M. Eastern time, so that voters throughout the nation, in all three time zones, could easily tune in after dinner.

His language was natural. As Brown reports, one study of the texts of the chats has revealed that almost three quarters of his words were to be found among the one thousand most commonly used in the English language. Where other presidents strived to be "presidential" when they spoke—dignified, exalted, and self-important—FDR tried to be friendly and simple. His easy familiarity seemed to warm both the radio set and his image at the same time.

During one Fireside Chat, on July 24, 1933, amid ninety-degree summer heat, FDR asked, on the air, "where's that glass of water?" "After a brief pause, during which he poured and drank the water," Brown recounts, "he told his national audience, 'my friends, it's very hot here in Washington tonight.'"

FDR realized, too, that speaking slowly was key. "While most radio orators were accustomed to speaking at one hundred and seventy-five and two hundred words per minute, the president consistently [spoke] at a much slower one hundred and twenty words . . . and during crucial broadcasts, when he wanted to convey the gravity of the situation, he reduced his rate of speech to less than one hundred words per minute so that everyone could understand." His pronun-ciation was clear, although rich with his confident eastern establish-ment inflection. Faintly British in accent, he spoke with a cadence and rhythm that made his speech all the more magnetic. The CBS speech critic John Carlisle commented that his vocal cords "were more sensitive and more susceptible to the influence of emotions than the strings of a violin to the master's touch of the bow."

Eleanor Roosevelt also understood FDR's appeal—and how he used it. "My husband had the very remarkable ability to project his personality through his speeches and he certainly had the ability to put into understandable English even quite difficult thoughts. He gave his listeners a feeling of gracious familiarity and sincere con-cern."

The manager of the CBS Washington Bureau, Harry Butcher, coined the term *Fireside Chat* after Roosevelt aide Steve Early said, "the president likes to think of the audience as being a few people around his fireside."

Some fireside! Here is how Robert Brown describes the scene in the White House when the president spoke: "Standing on either side of the President's desk were [reporters] from NBC . . . and . . . CBS, each having a microphone on a platform in front of them. Aimed at FDR were four huge motion picture cameras. . . . Also pointed in the President's direction were . . . five cameras for still pictures. The room was cluttered with all sorts of portable electrical apparatuses and the floor was strewn with electric cables leading to the Presi-dent's desk. Lining the room facing the President . . . were twenty-five radio engineers and sound and still photographers." Yet his sec-retary of labor, Frances Perkins, explained that he would "picture in his mind the audience he was addressing, and then his face would smile and light up as if he were actually sitting on the front porch in the parlor with them."

The *New York Times* concurred: "His magnetic voice has vibrated in space to clarify for the people the meaning of bank reform, inflation, legal beer, mortgage relief, taxation, and many knotty problems."

FDR tried to space his talks so as not to bore the public. He wrote that the "one thing I dread is that my talks would be so frequent as to lose their effectiveness." Even the example of another great communicator from overseas gave him pause: "I ought not to appear oftener than once every five or six weeks. I am inclined to think that in England, Churchill, for a while, talked too much and I don't want to do that."

As one of Roosevelt's pet projects—the Rural Electrification Administration—brought power to America's farms and hamlets, more and more people heard their president. For a typical chat, such as the one he gave on September 11, 1941, fifty-four million Americans tuned in—73 percent of the adult population.

In seizing the venue of broadcast radio for his speeches, FDR had created a "one man revolution of modern oratory," in the words of *Broadcasting*. While William Jennings Bryan and Daniel Webster were the "speech-making heroes of the nineteenth century, FDR is the oratorical model of the twentieth."

And as Roosevelt spoke, the nation listened. As his message reached into every home, voter participation in elections rose by 87 percent from 1920 to 1940, even though the U.S. population grew by only 25 percent during the same period. Of course, part of the reason for the increased voter interest was that the 1930s and early 1940s were desperate times, and vital issues were at stake. But it's impossible not to believe that the unprecedented reach of FDR's voice helped contribute to the increase. The voters were connecting with their President.

The banking speech was the first of four Fireside Chats during Roosevelt's first year in office. Half a million wrote letters and telegrams like this one: "My wife and I are two typical American citizens. . . . We were sitting in front of the fireplace . . . when your voice came to us from the White House. . . . We went into the next room and got in front of the radio, and for twenty minutes listened to you as though you were visiting us in our house and talking over our problems, and the problems of our friends and neighbors. One cannot listen to your frank, straightforward discussion . . . without being inspired with faith in your intentions, and hopes for your success."

One letter read, "You called us 'friends' over the air and I hope you will not consider me presumptuous when I sign myself, Your friend." Another thanked FDR for bringing to the presidency "the real genuine human touch that everybody has waited for so long."

Roosevelt's second Fireside Chat, two months later, explained the various New Deal measures he had pushed through Congress in his first sixty days. He gushed about the Conservation Civilian Corps, designed to put 2.5 million unemployed young Americans to work preserving the natural environment. He explained how the Federal Emergency Relief Act, the Tennessee Valley Authority (TVA), the Agricultural Adjustment Act (AAA), and the National Industrial Relations Act (NIRA) would reform the economy and fight the Depression.

Defending his programs against the charge that they were socialist, he articulated a new vision of government's role in American business: "It is wholly wrong to call the measures that we have taken, government control of farming, control of industry, and control of transportation. It is rather a partnership—a partnership between government farming industry and transportation."

Two months later, he used a third chat to enlist the cooperation of farmers, workers, shopkeepers, and manufacturers in following the voluntary guidelines set for wages, hours, and business practices in each industry by the NRA—the National Recovery Administration. To get them to accept these suggested rules, FDR once again turned to his friend the microphone. His speech was greeted with an outpouring of support, and the agency was successfully launched. Soon NRA stickers with their trademark eagle were in shops and stores throughout the land.

As opposition to FDR's reforms mounted, he used a Fireside Chat on June 28, 1934, to lash out at his critics and to ask voters, "Are you better off than you were last year? Are your debts less burdensome? Is your bank account more secure? Are your work conditions better? Is your faith in your own individual future more firmly grounded?" The answers helped to drive his critics from the stage as the congressional elections of 1934 approached. Almost fifty years later, Ronald Reagan would repeat almost the same questions in the 1980 presidential debates.

During Roosevelt's 1936 reelection campaign, radio was front and center in his arsenal. One month before election day, he used radio to accuse the Republicans of having brought the country "hear-nothing, see-nothing, do-nothing government" during their years in power. Roosevelt won by one of the biggest landslides in history, carrying all but two states and defeating the Republican Alfred Landon by eleven million votes. Democrats carried the House by 331 to 104, and the Senate by 76 to 20. As House Speaker Sam Rayburn told Roosevelt, "It was your nationwide radio speeches in 1936 which carried forty-six out of forty-eight states."

As he began his second term, Roosevelt made his first mistake. Determined to fight back against the Supreme Court, which had issued a constant stream of decisions invalidating key aspects of his New Deal program, he announced a plan to pack the court. He wanted to add positions for six new judges, claiming that workload on the bench had grown overwhelming and that the current judges needed help. To sell this radical proposal he should perhaps have turned to his best weapon, the radio. He chose, instead, to announce his court-packing plan at a press conference on February 5, 1937. Forgetting his own playbook, he spoke only to the press, failing to take to the radio to explain himself to America.

It was a gambit that didn't succeed. Indeed, his failure to use a Fireside Chat to rally public support in this instance is the exception that proves the rule of radio's effectiveness. From the start the press spun FDR's plan as underhanded and dictatorial, and congressional support ebbed away. By the end of February, Gallup polls showed that public support for FDR's plan had fallen below 40 percent. By the time Roosevelt relented and took his pitch for the plan to the radio audience almost a month later, it was too late. The plan died a quiet death in Congress; Sam Rosenman lamented that the episode "need not have been a disaster if it had been presented to the people correctly, instead of the way it was."

Learning from his mistake, Roosevelt took great care to use radio speeches as America's attention turned to the growing menace of war in Europe. His Fireside Chats increased public awareness of the threat posed by Nazi Germany and Imperial Japan. Fighting America's historic isolationism and deep disillusionment after World War I, Roosevelt sought to rally Americans to an understanding of the dangers at hand.

In six Fireside Chats between September 3, 1939, and October 27, 1941, Roosevelt attacked the growth of fascism in Europe. For each step in his gradual escalation of American involvement and preparedness, the president delivered a Fireside Chat to explain its necessity, and to catalyze support. People often wonder what the proper balance is between political leadership and sensitivity to public opinion. FDR's skill in leading his nation out of isolationism and into World War II through a series of brilliant Fireside Chats illustrates how to lead public opinion rather than just limp along after it.

Going to war is, of course, the most intrusive sacrifice a government can ask of its people. It involves not just the sacrifice of money, but the loss of lives. Radio's intimacy granted FDR a virtual place at the family dinner table, allowing him to participate directly in each household's discussion about the war, and to guide them, household by household, toward understanding why so great a commitment was required.

As soon as Germany invaded Poland on September 3, 1939, starting World War II, FDR took to the airwaves. For months he had been lobbying Congress to change the Neutrality Act that stopped him from selling arms to the Allies. Mindful of how the sinking of American merchant ships had catapulted the nation into World War I, Congress had prohibited the sale of weapons to belligerents in wartime. Three hours after Britain and France declared war, FDR went on the air to urge that the act be modified to allow nations to buy arms from the United States as long as they paid cash and took them home in their own ships—that is, on a "cash-and-carry" basis.

In his speech to an apprehensive nation, Roosevelt made clear where his heart lay as the war began. "Until 4:30 this morning," the president said, "I had hoped against hope that some miracle could prevent another devastating war in Europe and bring to an end the invasion of Poland by Germany. . . . Passionately though we may desire detachment, we are forced to realize that every word that comes through the air, every ship that sails the seas, every battle that is fought, does affect our American future. . . . The nation will remain a neutral nation, but I cannot ask that every American remain neutral in thought as well. Even a neutral has the right to take into account the facts. Even a neutral cannot be asked to close his conscience."

Calling Congress into special session, Roosevelt kept up the pressure for a change in the Neutrality Act with another radio speech. Before the broadcast, *Fortune* said that 50 percent favored repeal of the neutrality clauses; by September 21 it had risen to 70 percent. On November 4, a revised Neutrality Act was passed.

After the Nazi blitzkrieg rolled over Scandinavia and the Low Countries, FDR used radio again, this time to warn that France would soon be Hitler's next victim. On May 26, 1940, he told fearful Americans that the United States would not be caught flat-footed in facing Nazi aggression. As Roosevelt spoke, attendance at New York City movies and theaters dropped by 80 percent as listeners huddled anxiously around their radio sets at home to catch the president's words. "There are many among us who in the past have closed their eyes to events abroad—because they believed in utter good faith what some of their fellow Americans told them—that what was taking place in Europe was none of our business; that no matter what happened over there, the U.S. could always pursue its peaceful and unique course in the world." Now, he indicated, "it is our purpose not only to speed up production, but to increase the total facilities of the nation in such a way that they can be further enlarged to meet emergencies of the future." Pledging to ask Congress for larger defense appropriations, he declared, "It is my resolve and yours to . . . meet the present emergency."

Later in 1940, Roosevelt again gave a Fireside Chat to prepare the nation for the resumption of the military draft. Avoiding the odious words *draft* and *conscription,* he explained the need to augment our "peacetime army" that exists "for one purpose only: the defense of our freedom." The bill passed by one vote in the House, and by November 1 there were 1.2 million new recruits and 800,000 reservists added to the army.

The looming election of 1940 intervened as Roosevelt was trying to edge the United States toward intervention in the war. Ever since George Washington had retired after two terms in office, no president had ever served a third term. (Theodore Roosevelt had tried and lost, but there had been an interval between his two terms and his attempt at a third.)

Roosevelt made an elaborate show of not wanting to run, but ultimately agreed to be a candidate, telling the nation "My conscience will not let me turn my back on a call to service."

The key issue, of course, was U.S. involvement in the war. Republican nominee Wendell Willkie challenged Roosevelt, saying "if his promise to keep our boys out of foreign wars is no better than his promise to balance the budget, they're already on the transports." But on October 30, 1940, FDR countered, saying in a radio address "I have said this before and I shall say it again and again and again. Your boys are not going to be sent into any foreign wars."

Roosevelt surged to a 449–82 victory in the electoral vote, but the popular victory was closer.

Broadcasting wrote that "FDR's astute use of radio probably accounted as much as any other tangible factor for [his] reelection." *Variety* concurred: "His election for a third term demonstrated as nothing else could, the power of American radio. More than a political contest, the 1940 election was a battle between newspapers and radio to test which medium exercised the greatest influence on the American public. When the papers lined up almost 90% solidly against the third term, Roosevelt took his case directly to the people by the airwaves."

Meanwhile, with the fall of France, the British position became increasingly desperate. No longer could England afford to pay for the arms she needed to sustain the battle against Hitler. Roosevelt, in a daring move, proposed a program he called Lend Lease, designed to provide war materials and defer any payment, until after the war was over. Once again, FDR used radio to sell his program. In a homey metaphor he likened the Lend Lease program to lending his neighbor his garden hose to put out a fire. "You would not haggle about wanting fifteen dollars for the hose. You would tell your neighbor to take the hose and put out the flames and give the hose back later." Splendidly disregarding the reality—that tanks, planes, boats, and bullets weren't always so easily returned after combat—Roosevelt sold his generous program with the most irresistible of images. As

Rosenman observed, "as complicated and grandiose a scheme as Lend Lease turned out to be . . . it could never have been more simply or effectively placed before the American people."

Two weeks later, Roosevelt continued his campaign to make America an "arsenal of democracy," with a dramatic appeal to history: "Never before since Jamestown and Plymouth Rock has our American civilization been in such danger as now. . . . The Nazi masters of Germany have made it clear that they intend not only to dominate all life and thought in their own country, but also to enslave the whole of Europe, and then to use the resources of Europe to dominate the rest of the world."

Fifty-nine percent of the American people heard the speech over the radio—and they agreed with their president by 68 to 28 percent. Congress followed suit, approving Lend Lease in March of 1941.

On September 4 of that year, the United States moved closer to war when one of its destroyers, the U.S.S. *Greer,* was involved in a clash with a German submarine. FDR discussed the incident in his Fireside Chat one week later, giving a somewhat revisionist account of the incident. He said that the Germans had "initiated two unwarranted torpedo attacks" on the *Greer.* "The German sub fired upon this American destroyer without warning and with deliberate design to sink her. . . . Normal practices of diplomacy . . . are of no possible use in dealing with international outlaws who sink our ships and kill our citizens. . . . When you see a rattlesnake poised to strike, you do not wait until he struck before you crush him. These Nazi subs and raiders are the rattlesnakes of the Atlantic. . . . Their very presence in any waters which America deems vital to its defense constitutes an attack. . . . It is the time for action now."

Step by step, Roosevelt was using the radio to help listeners come to terms with the reality—that America's involvement in World War II was inevitable and necessary.

When the Japanese attack on Pearl Harbor on December 7, 1941, forced America into the war, FDR no longer needed to fight the isolationists. But he still had plenty of defeatists on his hands, who quailed in the face of Nazi and Japanese aggression. In a series of Fireside Chats—four in 1942, four in 1943, three in 1944—Roosevelt bolstered American morale, stimulated the hard work that underlay our industrial production, and smoothed over the inevitable labor problems that could have derailed production of the weapons of war.

Most important, he continued to report, to persuade, and to guide the American public—his "friends"—through the dark days of the war. FDR waged World War II as surely by radio as he did by military might.

JOHN F. KENNEDY USES TELEVISION TO WIN THE UNWINNABLE ELECTION . . . AND RICHARD NIXON USES IT TO LOSE THE UNLOSABLE ONE

If the message of radio is intimacy, what is the message of television? This question lay at the core of the Kennedy-Nixon contest of 1960.

John F. Kennedy understood that television was the medium to showcase his glamour and charisma, wit and intelligence. He realized that the simple difference between radio and television was that one had pictures and the other did not. And he spared no effort to make those visuals as attractive and dynamic as possible, imparting a carefully scripted but subtle message about himself and his candidacy.

Nixon, on the other hand, saw television as just another form of radio—a medium where words, not visual images, mattered the most. He totally missed the point of television. Like a movie actor from the old days of silent pictures, lost in the new world of talkies, he groped to understand how to use television, and failed miserably

The election hung on the difference.

Here were two candidates, both in their forties, both veterans of World War II, both with fourteen years of experience in public life. Either one would have been America's first G.I. Generation president. But only Kennedy perceived that the message of generational change was the key to victory, so, like an entrant in a beauty contest, he focused on looking and acting the part.

And Kennedy, as historian David Aberbach and others have commented, was the first politician who "willingly be[came] a media commodity." The golden boy of a political dynasty possessed of vitality and pride in equal measure, Kennedy came by his magnetic image not by happenstance, but through a lifetime of study, effort, and application. He and his consumingly ambitious father, Joseph P. Kennedy left nothing to chance.

Not even nature. John Kennedy may have had good genes, but he wasn't always as captivating as he later became: habitually underweight and chronically ill, as a young senator he was once described by none other than Lyndon Johnson as "malaria-ridden, yallah . . . sickly, sickly." Afflicted with continual maladies throughout his youth, Kennedy was periodically reduced to using crutches to walk, and spent months in the hospital for a spinal fusion in 1954. During that time, he was placed on the critical list twice, and was even given the last rites of his church as he lay near death. By the late 1950s he had seemingly recovered, and with the aid of "visits to Palm Beach and . . . a sun lamp," as confidant Ted Sorensen remembers, he was able to conceal his illness. Kennedy suffered from chronic Addison's disease, an inability of the adrenal glands to properly secrete cortisol. Even his brother Bobby joked "about the great risk a mosquito took in biting Jack Kennedy." By the 1960 presidential campaign the pale scarecrow of old had been transformed into a hale and hearty Greek god. What made the difference? Cortisone, a steroid he took to control his disease. The medication "transformed" Kennedy's face—and his political future along with it.

Even though the Kennedy family was reared amid the old-line machine politics of Boston, they understood, before most others, that modern politics was all about media—and, increasingly, about television. Father Joe, himself once a major Hollywood producer, had friends throughout the print, radio, and television industries— and his son inherited the older man's faith in the media and media experts. As historian David Burner notes, television "enabled Kennedy to present himself less as a politician and more as an independent force for moral reformation." Lavishly funded by his wealthy father, the Kennedy family hired the best experts to guide Jack through the shoals of the media.

When Kennedy ran for the Senate in 1952, he was one of the first American politicians to use television advertisements to promote his candidacy. Combining jingles with attractive footage of the candidate shaking hands, visiting factories, and talking with young people, the ads were the precursor of modern political advertising.

Kennedy took part in screen tests with different backdrops wearing different suits. He was careful to be constantly photographed in action playing touch football and going sailing to emphasize his youth, his vitality. Even when he was still sickly, Jack Kennedy always made sure he *appeared* the picture of perfect health.

After trying unsuccessfully to capture the vice presidential nomination on the ill-fated 1956 Democratic ticket, Kennedy assumed the role of spokesman for the party, introducing presidential nominee Adlai Stevenson at the convention. The speech—and Kennedy himself—got rave reviews. This was the moment the *New York Times* called him a "movie star." Jack Kennedy, the young senator, began to merge into JFK the national presence, in wide demand for appearances on behalf of his party, and an obvious candidate for a future presidential run.

To help burnish his image, Kennedy also wrote—or had written for him—the best-selling book *Profiles in Courage,* an account of singular acts of political bravery throughout American history. Taking to television to promote the book, Kennedy became an increasingly familiar and comfortable media figure. Hollywood director Dore Schary even selected JFK to narrate a film history of the Democratic party. As David Burne notes, "What better narrator than the author of a bestseller on political integrity, whose effective television presence had become increasingly noticeable?"

When Kennedy announced his presidential candidacy in 1960, he boldly indicated his intention to enter key Democratic primaries—a risky and, at the time, unusual step. Determined to prove that a Catholic could win votes even in Protestant states like West Virginia, he campaigned aggressively, cleverly making extensive use of television advertising to introduce John Kennedy to the voters in the way that he wanted the voters to know him. Once again his ads featured a jingle—but this time it was sung by none other than Frank Sinatra to the tune of the hit song "High Hopes."

The Kennedy campaign also hired five media advisers to help orchestrate the 1960 race. Three of them—Charles Guggenheim, Tony Schwartz, and David Sawyer—would become the most prominent media creators in American politics throughout the next two decades.

The campaign ads were unorthodox. Some were on location, another major innovation, which lent the ads a feeling of movement and energy. And Kennedy's photogenic family appeared in the ads, another first in American politics.

Reviewing the political landscape as it appeared at the start of the 1960 campaign, it's clear that Kennedy's appreciation of the importance of the media was one of the few undeniable advantages he brought to the race. Marshall McLuhan noted early that TV would always attract "the true Narcissus style of one hypnotized by the amputation and extension of his own being in a new technical form," and Kennedy, the self-mythologizer, clearly reveled in the ability to project his family's sparkling surface onto TV screens all over the country. He "used the medium with the same effectiveness that Roosevelt had learned to achieve by radio. With TV, Kennedy found it natural to involve the nation in the office of the Presidency, both as an operation and as an image."

And yet the irony is that Kennedy was going up against a man who many believed had mastered that medium already: Vice President Richard M. Nixon. By any conventional measure, Nixon should never have lost the presidential race of 1960. It took the combination of his opponent's skillful wielding of charisma—and a series of extraordinary miscalculations on Nixon's own part—to throw the election to Kennedy.

Nixon was coming off eight years as the vice president of possibly the best-liked man ever to occupy the White House. Widely celebrated as one of the heroes of World War II, Dwight D. Eisenhower, with his captivating grin, had held America in his thrall for two terms. Even after he almost died from a heart attack, voters reelected him overwhelmingly. Despite three recessions, he left office with as close to unanimous approval as any postwar president has come.

John F. Kennedy, meanwhile, was an underdog from the start, for one fundamental reason: He was only the second Catholic to run for president. The first, cigar-chomping New York governor Alfred E. Smith, had been annihilated by Herbert Hoover in 1928. With anti-Catholic prejudice rampant in the southern and border states—all of which any Democrat would need to carry to be elected—Kennedy's chances looked slim to none.

But then Nixon changed everything by agreeing to appear in four televised presidential debates—the first ever—with his telegenic rival. It was a bold step, and one that Eisenhower advised against. But Nixon was a confident debater. And if there was one American politician who could fairly claim that television was his medium—at least before 1960—it was Nixon.

After a career of hunting communists and spies, Richard Nixon had been chosen by Republican chieftains to run with Eisenhower on the 1952 presidential ticket. The result was a perfectly balanced ticket. Nixon was young; Eisenhower was old. Nixon was from California; Eisenhower lived in Pennsylvania. Nixon was a conservative; Ike was a moderate. Nixon had been a party warhorse for years, while the former general was a political ingenue.

Shortly after his nomination, though, scandal had threatened to make Nixon the first vice presidential candidate to be forced to resign after his nomination. Just as the campaign was getting under way, the *New York Post* ran the headline, "SECRET NIXON FUND! SECRET RICH MEN'S TRUST FUND KEEPS NIXON IN STYLE FAR BEYOND HIS SALARY." The issue became an embarrassment to the GOP—and especially Eisenhower himself. Proud of the reputation he had maintained throughout his long career of integrity and probity, Ike let it be known that he was thinking of dropping Nixon from the ticket. He asked Thomas E. Dewey, the two-time GOP standard-bearer, to phone Nixon to try to persuade him to leave the ticket.

Desperate to contain the damage, Nixon began casting about for a way to force Eisenhower to let him stay—and decided to gamble his career on a national television and radio address.

For twenty-seven hours, Senator Nixon worked on his speech at his suite at the Ambassador Hotel in Los Angeles, stopping for only an hour to take a swim in the hotel pool. It was classic Nixon—the lonely workhorse, throwing himself into death-march mode to bulldog his way through a crisis.

Nixon's speech was as well-rehearsed as any in American politics, and it was painstakingly orchestrated in the studio by media expert Ted Rogers, previously a production supervisor for *The Lone Ranger* and *Mystery Theater.* "The TV experts had Nixon sit at the table, then stand next to it and then do the same thing again and again . . . practice posing with his right hand on the table, then with his left hand in a trouser pocket," as one account recalled. Pat Nixon "was carefully coached to keep a relaxed pose in her chair, her head turned at a certain angle, her face arched in a close-mouthed smile."

Nixon greeted his audience in a gray suit and a dark tie, as fifty-eight million people watched—the largest TV audience in history. Though fighting for his political life, Nixon spoke with strength and assurance. He described his modest upbringing, defended the fund, and told America that "Pat doesn't have a mink coat. But she does have a respectable Republican cloth coat. And I always tell her that she'd look good in anything."

The performance reached its sudsy climax when Nixon, consciously imitating FDR's use of his dog Fala in his 1944 campaign, revealed that he, too, had received a canine gift in the campaign. "[A] Texas supporter," he told the audience, had given his daughters, Tricia and Julie, "a little cocker spaniel dog . . . black and white spotted." The kids named it Checkers. "Regardless of what they say about it, we're gonna keep it," the candidate told America, addressing the camera directly in a moment of extraordinary television. Nixon later said that he felt "that surge of confidence that comes when a good speech has been well prepared," even as he waited to learn whether his gambit had worked.

America was touched.

One Nixon advisor, worried that the speech was too corny, learned the truth "when I saw the elevator operator crying." To the viewers, Nixon seemed to be "a figure from a Frank Capra movie, a 'Mr. Smith' who had gone to Washington and found himself contending with all the problems that the Mr. Smiths of America could recognize."

All over America, phone switchboards were buzzing with support for the embattled candidate. Aides told Nixon "the telephones are going crazy; everybody's in your corner!" The Republican National Committee "in Washington alone reported the receipt of 300,000 letters and telegrams signed by more than a million people" in support of Nixon. Polls found strong support for keeping Nixon on the ticket. As historian Herbert Parmet writes, "The Checkers speech as much as the [Alger] Hiss affair established Nixon as a political figure."

Eisenhower ended Nixon's agony by flashing his infectious grin, "hand outstretched," saying "you're my boy."

Nixon, who saved lessons on how to deal with scandal like a camel stores water, said he learned from the fund crisis that "it isn't what the facts are but what they appear to be that counts when you are under fire in a political campaign." For him, television was now the forum for effectively and successfully communicating to the nation.

But Nixon's decision to debate Kennedy on TV stemmed not just from his confidence about television; it also had a great deal to do with his opinion of himself as a debater. A member of a championship debating team in school, as a politician Nixon had sealed his reputation for being able to handle foreign affairs through a bizarre encounter with Soviet leader Nikita Khrushchev, that became known as the "Kitchen Debate."

Touring a United States–sponsored exhibit on American family life in Moscow with the Russian dictator, Nixon was amazed when Khrushchev began debating political theory with him in front of the press. Khrushchev immediately took the lead as the pair toured a model television studio. Dominating the conversation, Khrushchev told the American vice president "you don't know anything about communism except fear of it."

As the tour continued into the kitchen exhibit, Nixon accused Khrushchev of filibustering. "You do all the talking and you do not let anyone talk."

Those who read the transcript of the "debate" were struck by how the Russian had initially gained the upper hand. But Nixon came out on top because of photographs appearing in the press that depicted him as "the man who stood up to Khrushchev."

Buoyed by his success at using television and "winning" debates, Nixon jumped at Kennedy's proposal for televised debates. Everyone knew the race was close. When a September Gallup Poll showed Nixon ahead by only a single percentage point, it became clear that the debates would be pivotal.

Why did Nixon, the sitting vice president, agree to debate the little-known senator from Massachusetts? Ted Sorensen himself wondered about the decision: "[Nixon] had no reason to help build up an audience for Kennedy."

But historian Earl Mazo thinks Nixon "was convinced he could win. . . . He had stayed on the ticket in 1952 by his effective use of television; he reached his highest popularity after the 'Kitchen Debate' with Khrushchev in 1959. Now by combining debating and television" he thought he could use his skill at winning arguments to win votes.

Paranoia also played its role—as it always did with Nixon. Suspicious of the liberal eastern media, Nixon was convinced he would never get a fair break from the print press. To Nixon, television was the great equalizer. In front of the camera, he thought he could do what he had always done best—speak directly to voters and win their trust.

But Nixon's visage was better suited to Iowa cornfields than it was to television. His "style was a simple earthy one and could only be appreciated in little groups in [small] towns." In person, White observed, the forty-seven-year-old Nixon appeared "attractively slim . . . lithe . . . a fine and healthy American" with a face that had "a clean, masculine quality." But his TV image was a different story. "On television, the deep eye wells and the heavy brows cast shadow on the face and it glowered on the screen darkly; [and] when he became rhetorically indignant, the television showed ferocity."

Having taken a questionable risk in even agreeing to the debate, Nixon aggravated the mistake by refusing to work with television experts as he readied himself for the showdown. Eisenhower suggested that Nixon use television producer Robert Montgomery, "an expert in lighting, makeup, and other aspects of television," to prepare for the debate. Confident that he knew better, Nixon refused.

As they prepared for the debate, Kennedy did everything right, whereas Nixon made mistake after mistake.

- Kennedy was briefed again and again in the days before the debate; Nixon never prepared properly.

- Kennedy was rested and tanned for the debate; Nixon looked "white and pasty."
- Kennedy wore a dark suit to offer a contrast with the backdrop; Nixon's light gray suit blended in, creating a "fuzzed outline."
- Kennedy wore makeup and was clean-shaven; Nixon refused cosmetics and tried to cover his beard with a product called Lazy Shave that didn't work.
- Kennedy looked into the camera, addressing voters directly; Nixon recalled his college debating style and glared at Kennedy, ignoring the cameras.
- Kennedy looked cool and in command; Nixon sweated beneath the camera lights.
- Kennedy kept his hands folded in his lap and his legs crossed; Nixon fidgeted and kept his hand awkwardly pinned to his leg.

Kennedy looked like a teenage girl's dream date. Nixon looked like the angry father who meets his daughter at the door when she comes home late.

Kennedy won. Nixon lost. Those who heard the debate on radio felt Nixon had done better, but the vast majority who saw it on TV sided with Kennedy.

The question is: why did Nixon screw it up?

He was a man of the 1950s. As McLuhan noted, television is a cool medium. But there was nothing cool about the Richard Nixon of the Eisenhower era. He had made his career prosecuting communists and savaging his Democratic opponents.

He had found success by baiting Democrats with catch phrases like "Dean Acheson's Cowardly College of Communist Containment." His very name had become a synonym for confrontation, accusation, and the hard exchanges of politics.

He felt he would defeat Kennedy in the debates because of his intimate familiarity with issues and his mastery of political discourse. He had gone up against the best of them—Truman, Acheson, Stevenson—and prevailed in his long career.

Appearances—did they matter? Should they count?

Reviled by the liberal intellectual establishment for his performance in the Checkers speech, Nixon had come to be sensitive to the idea that he was skating by on style and cornpone sentiment. He wanted to be seen as a candidate of substance.

The Checkers episode had left a "deep scar," he confessed, "which was never to heal completely." In its aftermath, Nixon had come to develop a snobbery about television, complaining that it emphasized "showmanship" over "statesmanship." He didn't get it.

Indeed, Nixon's later accounts of his 1960 debate performance suggest that he took a certain satisfaction in the fact that he lost over appearances and not over substance. Perhaps he was more eager to demonstrate his command of the issues than he was to win the debate.

In his memoirs he sniffed: "It is a devastating commentary on the nature of television as a political medium that what hurt me most in the first debate was not the substance of the encounter between Kennedy and me, but the disadvantageous contrast in our physical appearances." In his earlier book *Six Crises*, Nixon seemed almost proud of his failure: "I had concentrated too much on substance and not enough on appearance."

The vanity and private conflicts of public men and women so often bring victory or lead to defeat. Who could have known that, at the back of Nixon's mind as he faced Kennedy that night, lay the wounds of an ego injured and battered by the criticism of his Checkers speech? This graduate of an obscure western college never felt accepted by the eastern elite. How much of Nixon's bad-boy affect was due to this feeling of rejection by the establishment? To what extent was his paranoia, which ultimately consumed his presidency, due to his resentment at the contempt with which the nation's intellectual elite treated him?

Now, pitted against Kennedy in the battle of his life, he chose not to wear makeup, not to get rest, not to dress in the right suit, not to speak to the camera, not to do anything right. Why? Was he perhaps marching to the beat of an inner drummer no one but he could hear?

Listen for the sound of that drummer in Nixon's press secretary Herb Klein's explanation, given years later and recounted in Christopher Matthews' *Kennedy and Nixon*. Klein said Nixon had "heard Jack Kennedy's derision of Hubert Humphrey for wearing heavy makeup in their joint television appearances during the Wisconsin primary campaign. 'To Nixon, this made it look like he lacked macho [if he wore makeup] and Nixon was a very macho man.'"

Insecure about his appearance, Nixon let his *opponent* determine whether or not he used makeup! Perhaps at some level Nixon simply didn't want to win if victory could only be bought with the same kind of derision he suffered after the Checkers speech.

Kennedy, on the other hand, gloried in the stylistic contrast that television made possible. As he told reporter Rowland Evans, "for the first time since the Greek city-states practiced their form of democracy, it brings us within reach of that ideal where every voter has a chance to measure the candidate himself." For him, a picture was better than a thousand words. "Television," he said, "gives people a chance to look at their candidates close up and close to the bone."

That was easy for the Greek God Jack Kennedy to say. But for Nixon, the contrast was less than flattering.

In a way, though, Kennedy had a point. Nixon's later commentary about the debates suggests that his real problem was a kind of essentially antidemocratic arrogance. Nixon's self-involved disdain for the showmanship of TV seems to have blinded him to the medium's role as a direct channel to the constituency he was trying to reach. Instead of looking out at them, he looked in the mirror and brooded over his self-image.

Nixon ignored television for the rest of the campaign. His key advisors worked under the assumption that newspapers were the only medium worth courting. Gene Wyckoff, a television producer who worked for Nixon in the 1960 campaign, later wrote that Nixon "did not want the Madison Avenue stigma associated with him."

Even when Wyckoff produced two thirty-minute campaign programs, Nixon wouldn't run them, foolishly "relegat[ing] such television effort to a low priority in his campaign." Ted Rogers, who had coached Nixon in the Checkers Debate, produced a film that aired in California, but the campaign canceled plans to run it nationally, choosing instead to televise an election eve telethon, once again resisting the idea of a well-produced packaged television program.

Kennedy had no such problem. He loved the camera and it loved him. As he took the oath of office as America's thirty-fifth president, he said "the torch has been passed to a new generation of Americans." And he looked the part.

LYNDON JOHNSON RUNS NEGATIVE ADS . . . AND TRANSFORMS POLITICS

Just as Roosevelt used radio and Kennedy relied on television to move their careers, Lyndon Baines Johnson was the first candidate for president to use paid advertising as the core of his political campaign as he demolished Barry Goldwater in 1964.

If the "message" of radio is intimacy and that of television is glamour, what is the essential message of political television advertising? Republican advertising guru Bob Goodman once told me, "I deal in four things and four things only: love, hope, hate and fear." As Goodman realized, the essential "message" of television advertising is emotion.

Only thirty seconds long, the average political television ad must pack an emotional punch if it is going to be remembered. Lyndon Johnson was the first candidate to grasp this essential point.

Until 1964, paid television advertising had little impact on presidential elections. Both Eisenhower and Stevenson gave long televised speeches in 1956, the Kennedy campaign ran jingles in 1960, and Nixon held an election-eve telethon the same year. But these ads were sideshows.

In 1964, all that changed. In the course of one election season, television advertising moved to the center stage of American politics.

As often happens when a new technology is unearthed, it sweeps all before it. The negative advertising of the Johnson campaign so stigmatized Barry Goldwater that he remains, to this day, a symbol of trigger-happy policy and right-wing extremism.

Televised political advertising had started in the 1952 Eisenhower-Stevenson campaign. By then, nineteen million homes had television sets, and the medium was beginning to shape America. (By contrast, in the campaign of 1948, Thomas E. Dewey had been asked if he would use TV ads; "I don't think it would be dignified," he replied.)

Eisenhower's ads, produced by the Madison Avenue firm BBD&O, were a loose variation on the newsreels commonly shown in movie theaters of the time, with Eisenhower smiling broadly as the Republicans pushed their slogan: "I Like Ike."

In 1960, of course, John F. Kennedy had pinned his high hopes on Frank Sinatra's exuberant jingle ads. Yet it was also the Kennedy campaign that introduced the negative ad, taking advantage of a famous gaffe of Dwight Eisenhower's. Asked at a press conference whether he could recall an important decision on which he had consulted his vice president, Richard Nixon, Eisenhower had replied, "If you give me a week, I might think of one." It was the perfect counter to Nixon's slogan "Experience Counts," and Kennedy's team turned the moment into a withering ad.

But before 1964, the real race was still a matter of personal appearances, speeches, conventions, and, in 1960, the televised debates. It wasn't until Lyndon B. Johnson ran for the presidency that the negative ad became a factor in a presidential campaign. Indeed, that year negative ads *were* the campaign.

It wasn't that the Democrats spent more on TV advertising than the Republicans. In fact, Johnson spent $4.7 million on television, while Goldwater paid out $6.4 million. The difference was that Lyndon Johnson's people took a new tack: a scorched-earth barrage of negative advertising. Goldwater bought a lot of television time, but his spots lacked the aggressive and hard-hitting impact of the Johnson campaign.

Johnson's advisors grasped that the world was changing, and that television was ready for more than just bland positive messages. "Goldwater had followed outmoded advertising tactics at the moment the art was changing," Edwin Diamond and Stephen Bates

note in *The Spot,* their authoritative history of political advertising
on TV.

Through the lens of history, it wouldn't seem that Lyndon Johnson faced a difficult task in winning the election in 1964. Held less
than a year after President John F. Kennedy was slain, the contest
was inevitably colored by the enormous emotional outpouring of
grief that followed Kennedy's death, a factor that swung in Johnson's favor. Moreover, the Texan had amassed an unquestionable
record of accomplishment in the few months he had been in
office.

Kennedy had called upon his new generation in 1960 with the
words "Let us begin." Upon assuming Kennedy's mantle after the
assassination, Johnson implored Congress to carry forward
Kennedy's legacy, saying "let us continue." Calling on the House and
Senate to memorialize Kennedy by passing the sweeping civil-rights
bill the martyred president had proposed, Johnson worked day and
night to pass the measure. When he succeeded, it was an achievement of truly historic dimensions—and a dramatic display of Johnson's political might.

But Lyndon Johnson did not like elections. Elected by a paper-
thin margin in his first race for the House, Johnson by most accounts
actually lost his Texas senatorial campaign—until a batch of
"delayed" ballots mysteriously arrived from two staunchly Democratic and notoriously corrupt counties, giving Johnson a bare margin of victory.

So visceral was Johnson's fear of voters that he did not run in any
primaries as he sought the 1960 presidential nomination, pleading
that he was too busy doing the nation's work as Senate majority
leader to have time to campaign. (In later years, LBJ's terror of defeat
almost certainly motivated his withdrawal from the race in 1968.)

Whatever his private fears, what better way to avoid the scrutiny of
the voters than by focusing on the flaws of one's opponent? Johnson
gave this time-honored strategy of electoral jousting a new twist when
he decided to launch a series of negative missiles directly at Barry
Goldwater through the vehicle of TV advertising.

Johnson laid the basis for his negative campaign at his party's
national convention. Rather than devote the convention speeches to
a recitation of his own formidable accomplishments, he chose to ask

his newly anointed vice presidential nominee, the ebullient senator Hubert H. Humphrey of Minnesota, to savage Goldwater. Humphrey rose to the challenge, and Goldwater had given him plenty of ammunition: the Arizona conservative had been doing a thorough job of alienating the formerly ascendant moderate wing of his party in his drive for the nomination, leaving a hole broad enough for Humphrey to drive the whole Democratic machine through. Playing to moderate Republicans' disaffection for their new nominee, Humphrey called Goldwater "the temporary spokesman" of the Republican Party, charging that he was "out of tune with the great majority of his countrymen.

"Most Democrats and most Republicans in the United States Senate, for example," Humphrey said, "voted for the nuclear test ban treaty but not the temporary Republican spokesman.

"Most Democrats and Republicans in the Senate voted for an $11.5 billion tax cut for American citizens and American business, but not Senator Goldwater . . .

"Most Democrats and Republicans in the Senate—in fact four-fifths of the members of his own party—voted for the Civil Rights Act." Now the crowd was carrying Humphrey, chorusing along with him, *"But not Senator Goldwater!"*

On and on it went. With each peroration, the audience's enthusiasm, laughter, and sheer partisan joy grew.

But then Humphrey threw his best punch. In a speech to the Veterans of Foreign Wars, Goldwater had recently said that "small conventional nuclear weapons are no more powerful than the firepower you have faced on the battlefield. They simply come in a smaller package." The candidate speculated that the United States might use these "low level" nuclear weapons to defoliate the Vietnam battlefields to deny the Viet Cong cover. Somberly reminding the nation that "western civilization can be brought down in ruins in one hour," Humphrey now reminded his assembled party of the "compelling need for restraint in the use of the greatest power ever assembled by man." Goldwater, Humphrey warned, was guilty of "recklessness and irrationalism" and too often spoke with the "stridency and unrestrained passion of extreme and radical language." Humphrey concluded that "the American presidency is not a place for a man who is impetuous."

The "bomb" issue was born—one that would eventually help LBJ to destroy Goldwater once and for all.

Almost as soon as Humphrey sat down to the deafening applause of the convention audience, Johnson's advisors began to worry that his speech might fade from memory and that Goldwater might yet regain traction. Their worries were stoked by Goldwater's efforts to recover. One of his erstwhile Republican rivals, Governor William Scranton of Pennsylvania, convened a "unity conference" of Republican Party leaders in Hershey, Pennsylvania, on August 12, 1964. Party standard-bearers Dwight Eisenhower and Richard Nixon were in attendance as Goldwater promised, "I seek the support of no extremist."

Goldwater's repudiation of extremism left Johnson and the Democrats scrambling for a way to reignite the flame of suspicion Humphrey had lit at the convention. As then-Johnson advisor Bill Moyers recalled: "After the Republican convention, the president called me and said 'Barry [Goldwater] is making a headlong race for respectability. He's trying to shed himself of all of the convictions, beliefs, images and the extremism that had surrounded him all of his career in the Senate and characterized his speeches to the right wing . . . and he's trying to make himself respectable. What we've got to do is remind people of what Barry Goldwater was before he was nominated for president.' "

Moyers passed the command from on high to Johnson's TV advisors, and to Doyle Dane Bernbach (DDB), their ad agency. The birth of heavy-duty negative TV advertising was at hand.

Johnson aide Jack Valenti recalls telling Moyers and others on September 14, 1964: "If we are not careful, the Goldwater image will get all smoothed up to our detriment. Right now, the biggest asset we have is Goldwater's alleged instability in re atom and hydrogen bombs. We MUST NOT let this slip away."

Valenti urged that the campaign take the case against Goldwater onto television in paid ads. "Television ads," he said, could "be a heavy weapon for us . . . our real attack on Goldwater without involving the President." Johnson, anxious not to get his own hands dirty by attacking Goldwater himself, approved a hefty budget for television advertising.

Johnson's team of media advisors included Moyers, Valenti, and

Walter Jenkins. But his secret weapon, laboring away in a West 56th Street brownstone in Manhattan, was Tony Schwartz—a disciple of Marshall McLuhan, and a reclusive genius. As noted by Diamond and Bates, Schwartz was recruited for the Johnson campaign by DDB's Aaron Erlich, with whom Schwartz had worked on an American Airlines campaign. When Erlich showed Schwartz a picture of the president and asked, "Would you work for this product?" Schwartz—a dedicated Democrat—said that he would.

When the DDB people suggested using a missile countdown in a negative ad about Goldwater's trigger-happy attitude, Schwartz immediately thought of a recording he'd made in the 1950s of his nephew counting, incorrectly mixing up his numbers. An idea was born: Schwartz suggested juxtaposing the countdown with the sound of a little girl counting petals as she pulled them off a daisy.

The concept was brilliant and brutally effective. The final ad showed a little girl in a field pulling the petals from a daisy as she tries to count them: "one, two, three, four, five, seven, six, six, eight, nine, nine," she says in her innocent voice. When she reaches ten, a resounding male voice suddenly reverses the count: "Ten, nine, eight, seven, six, five, four, three, two, one." At zero comes a deafening roar, and the screen fills with the mushroom cloud of an atomic bomb.

Then one hears the voice of Lyndon Johnson: "These are the stakes—to make a world in which all of God's children can live, or to go into the darkness. We must either love each other, or we must die." A reassuring male voice concludes the ad: "Vote for President Johnson on November 3. The stakes are too high for you to stay home." The ad never mentioned Barry Goldwater, but everyone knew that it was about him. As Schwartz later wrote, "The commercial *evoked* a deep feeling in many people that Goldwater might actually use nuclear weapons. This mistrust was not in the *daisy* spot. It was in the people who viewed the commercial." Voters had read about Goldwater's consideration of using the bomb; now the ad tapped into that fear.

The ad, titled "Peace, Little Girl" and forever known as the "daisy spot," ran only once, on CBS' *Monday Night at the Movies* on September 7, 1964. Although it would air only that one time, Moyers recalls, "all three networks . . . picked up on the story and showed the commercial on their evening newscasts, giving the spot millions of dollars of free air time." The White House switchboard went nuts. Moyers told Johnson, "You got your point across." Johnson's delighted response: "You sure we ought to run it just once?"

How should Goldwater have handled the daisy spot? In a 2002 interview with Schwartz—who still spins miracles in his 56th Street office—he had a novel idea. "Goldwater should have said, 'I dedicate this campaign to banning nuclear war. So I will pay for this ad to air a second time, sponsored by my campaign.'" If Goldwater could have devised a way to use his own voice instead of Johnson's at the end of the commercial, it might have worked brilliantly. Had Goldwater simply gotten out of the way of the ad and not accepted that it was targeted at him, he could have ducked the punch deftly enough to make Muhammad Ali proud.

Instead, Goldwater stood up and took it on the chin. "Every time I saw that hideous Johnson TV commercial with the little girl, it saddened me to realize that all involved . . . valued political victory more than personal honesty." As Schwartz noted, "When people hear 'atomic weapons' they don't hear the word 'tactical.'" The idea of tactical nuclear weapons was almost meaningless, as Goldwater's staff had discovered. Years later, when Schwartz met Goldwater campaign director F. Clifton White, White described the impact of the ad on potential voters. No matter how much White would try to explain to wary listeners that Goldwater only envisioned using tactical nuclear weapons, as Schwartz recalled to Diamond and Bates, the person would respond, "Yeah, but we can't drop the bomb, Clif."

Schwartz had pioneered a new media strategy: the use of paid media, as the adman described it, "to put non-paid media in context." Political ads, he found, needn't provide a new message or new information; they could also serve to remind the viewer of what he already knows, and put it in a politically useful framework.

Schwartz also recognized TV's power to render familiar images larger than life; "experiences with television and radio stimuli," he noted, "are often more real than first-hand, face-to-face experiences." In his book *The Responsive Chord,* he told a story: "My wife loves to tell of a neighbor who was sitting in a park with her baby when another woman passed and commented, 'Oh, what a beautiful baby.' 'Yes,' replied the neighbor, 'but you should see his pictures.'"

The Johnson campaign didn't let up after the daisy ad. Ten days later, LBJ's media team came back with a second devastating blow.

Barry Goldwater had voted against the historic treaty Kennedy negotiated with Russia banning atmospheric nuclear tests, alienating the tens of millions of mothers and fathers who worried about the impact of nuclear fallout on foods—particularly in the milk their children drank.

In its second anti-Goldwater ad, another spot that ran only once, Johnson's team again featured a little girl—this one licking an ice cream cone. This time, a woman's voice-over—a first in political ads—asked: "Do you know what people used to do? They used to explode atomic bombs in the air. Now, children should have lots of vitamin A and calcium but they shouldn't have any strontium 90 or cesium 137. These things come from atomic bombs, and they're radioactive. They can make you die. Do you know what people finally did? They got together and signed a nuclear test ban treaty, and then the radioactive poisons started to go away. But now, there is a man who wants to be president of the United States, and he doesn't like this treaty. He fought against it. He wants to go on testing more bombs. His name is Barry Goldwater, and if he's elected, they might start testing all over again."

The Johnson offensive rolled on. Negative ad after negative ad destroyed all that was left of Barry Goldwater:

- One ad showed two hands tearing up a Social Security card while the announcer warned, "On at least seven occasions, Senator Barry Goldwater said that he would change the present Social Security system. But even his running mate William Miller admits that Senator Goldwater's voluntary plan would destroy your Social Security."
- A red phone rang in another commercial, as the announcer said "this particular phone only rings in a serious crisis." He then urges viewers to "leave it in the hands of a man who has proven himself responsible."
- Yet another ad quoted Goldwater as calling the atomic bomb "merely another weapon."
- One man-in-the-street ad featured Republicans criticizing Goldwater, while another quoted former Goldwater adver-

saries Republican governors Rockefeller, Scranton, and Romney knocking him.

Johnson punctuated his TV campaign by placing people at Gold-
water campaign rallies with derogatory placards reading "In Your
Heart, You Know He Might"—a play on the GOP slogan "In Your
Heart, You Know He's Right."

Such events were devised by the Five O'Clock Club, Johnson's
early version of the "war room," which met daily to plan the harass-
ment of Goldwater on the campaign trail. As Theodore White
recounted, "The offensive against Goldwater rolled almost by itself—
the nightly radio and television news spots; the organizational and
creative deviltry of the Five O'Clock Club; the independent counter-
offensive of Hubert Humphrey—all these held Goldwater in torment
on the issues of [the] bomb and Social Security, while the President
was free to spread balm and peace and friendship and promises all
around the nation."

Negative ads like these against a candidate like Goldwater would
have worked at any time, in any year. But the new technology ravaged
the GOP in 1964 as surely as the plague once decimated Europe.
And for the same reason—voters had no immunity. They had not
built up the cynicism and skepticism that now condition their
response to negative ads. If it was on television, they believed, it was
probably true.

By October, polls were showing that voters, by a margin of five to
one, thought that Goldwater was more likely to start a nuclear war
than Johnson. A quarter of those asked described him as "reckless."

The bomb issue put Goldwater on the defensive, haunting the
GOP campaign and sapping their will to launch a counterattack.
Goldwater became the first candidate to understand what one later
history of negative advertising concluded: That "the breadth of television's reach makes it difficult to dispel rumors or counteract the
effects of negative information." As Goldwater aide Denison Kitchel
later said, "My candidate had been branded a bomb dropper—and I
couldn't figure out how to lick it."

Goldwater became obsessed with the futile quest to prove that he
wasn't out to blow up the world, but his efforts may have done more
harm than good. When Charles Mohr of the *New York Times* moni-

tored a Goldwater speech in Indiana, he gave up keeping track after the candidate used such phrases as "holocaust," "push the button," and "atomic weapons" almost thirty times as he struggled to allay voter fears.

Goldwater even tried running a thirty-minute television show with former President Eisenhower to counter Johnson's thirty-second ads. Eisenhower obligingly called fears about Goldwater "tommyrot." But the show had little effect in the face of Johnson's offensive.

Having failed to rebut the Johnson attacks, Goldwater eventually went negative himself. A GOP ad featured Goldwater criticizing Johnson for moral decay, and asked, "What has happened to America?" Another ad attacked Johnson for being "so indecisive that he has no policy at all." But America hadn't yet developed any suspicions about Johnson; with no basis in a preexisting news story, Goldwater's attacks seemed ham-handed and artificial, a pale imitation of Johnson's carefully guided missiles.

Tony Schwartz likened good political commercials to the Rorschach ink blots used by psychiatrists to evaluate their patients. Just as shrinks hope their patients will read telling patterns into the blots, so Schwartz said the most effective ads don't "tell the viewer anything. They surface his feelings and provide a context for him to express these feelings."

The new medium worked. Lyndon Johnson won the 1964 election with 61 percent of the vote, while Goldwater carried only the deep South and his native Arizona. And in the process, American politics had changed forever.

Radio and TV, debates and advertisements: Each new advance in technology changes the rules of politics, and opens the door wider to the direct democracy of the Jeffersonian ideal. Whether the innovation brings the human voice into American living rooms, or enables two people to establish direct contact through computer technology, the effect is the same—to increase the involvement of the public in the political process and to raise the amount of information it receives. Politicians who exploit that new involvement tend to prosper; those who yearn nostalgically for the days when people knew less and learned less about politics, on the other hand, are prone to fall victim to the march of time.

MOBILIZING THE NATION IN TIMES OF CRISIS

Bombs fall on Afghanistan as this is written. President George W. Bush works to rally the American people to face terrorism even as the collapse of the twin towers of the World Trade Center replays itself endlessly in our mind's eye. How is he to mobilize America for the long, hard battle not just against Osama bin Laden and the Taliban, but against terrorist forces everywhere, a nihilistic and elusive foe?

Three examples offer compelling lessons from which we can—and must—learn. As a yardstick for success, none is better than the magnificent mobilization of Britain by Winston Churchill and of America by Franklin Roosevelt at the start of World War II. But the contrast of the failure of Lyndon Johnson to gird Americans for the decade-long war in Vietnam may teach us even more. By comparing Johnson's mistakes with FDR's and Churchill's brilliant successes, the student of history can learn much about the challenge of turning political power into military success—and vice versa.

And how is Bush doing? This is written in the opening days of 2002, as the Northern Alliance, led by U.S. advisors and bolstered by U.S. air power, have successfully driven the Taliban from Afghanistan. It is still early in the war on terrorism, but not too soon to give him high marks in emulating the Rooseveltian and Churchillian examples, and avoiding the Johnsonian mistakes.

In such times of crisis, history's lessons are clear.

MARK THE DEPARTURE FROM THE NORM

FDR and Churchill left no doubt in the minds of their people that they faced an entirely new situation as they were catapulted into World War II. No semblance of normal times remained. Johnson did the opposite; he blurred the changes the Vietnam war would bring, stressing continuity and concealing the massive U.S. commitment even as our soldiers were preparing to fight. With the attacks of September 11th of course, the world changed in an instant. And George W. Bush instinctively understood how to respond. By persistently stressing the nature of terrorism, Bush has done what he must to mark the start of a new era with different challenges and revised rules.

ACKNOWLEDGE THE MAGNITUDE OF THE TASK

Where Roosevelt and Churchill emphasized the difficulty of the task that lay ahead and the gravity of the sacrifice it would require, Johnson minimized them. He forced the nation to discover on its own the magnitude of the effort Vietnam would require, as the country drifted into a massive mobilization in an unfamiliar territory. By warning of the dangers of biological, chemical, and nuclear terrorism, Bush has honestly and frankly acknowledged the grave issues at stake for America in its new war. With no fear of telling the truth, he has laid the dangers out clearly to the American people. And, like Roosevelt and Churchill, he has become an admired leader with the overwhelming support of the country.

BEWARE OF FALSE GOOD NEWS

Worried lest their nations should prematurely relax their efforts and lose focus, the Allied leaders of World War II worked as hard to dampen facile optimism as they did to discourage defeatism. Even amid news of victory, they constantly pointed to the long road ahead. But the wisdom of that cautious strategy was lost on Lyndon Johnson. Instead of patience, he aggressively and repeatedly claimed victory was near. Each week he released a body count of enemy dead in Vietnam, and he seized on any glimmer of hope as evidence that the war was being won—until eventually people stopped believing him. As of this writing, Bush seems to be on the right track, as he warns of tough times ahead, and tempers the euphoria over the defeat of the Taliban and stresses the dangerous days of warfare that remain.

BRING CONGRESS INTO YOUR CONFIDENCE

Churchill cast aside party government altogether as he took power, forming a coalition with the opposition Labour Party. Roosevelt reached across the aisle and appointed Republicans to key positions such as secretary of war and secretary of the navy. But Johnson deluded Congress, just as he tried to fool the people. Lying about the origins of an incident in the Gulf of Tonkin, he tricked Congress into voting him power to wage war—and thereafter refused to consult with the legislative branch. Thus far, Bush has worked assiduously to bring Democratic congressional leaders into his inner circle, overcoming his fear of leaks to see that they are well informed and fully involved.

DON'T LIE

The Allied leaders of World War II grasped that the public would tolerate wartime secrecy and deception designed to throw off the enemy. But they never lied on basic questions. When they needed more troops, they said so. When adversity loomed, they admitted it. Johnson, however, based his entire conduct of the war on covering up the facts. Seeking not so much to fool the enemy as to deceive his own people, he duped the public by concealing new troop commitments and downplaying escalations. Bush has managed to avoid lying by straightforwardly explaining that he simply can't discuss much of his strategic planning without compromising the war effort—and in doing so has shown a level of trust in the people that Johnson never managed.

SHELVE DOMESTIC PRIORITIES WHEN FIGHTING A WAR

The New Deal stopped when Pearl Harbor was attacked. But during the escalation in Vietnam, Johnson doggedly insisted on pushing ahead with his Great Society domestic program, insisting that America could afford "guns and butter." His refusal to consider any more than a token tax increase was an error that would lead to a decade of inflation and economic stagnation. Bush faces the war on terror on the heels of having signed into law his cherished tax cut. While frankly embracing the necessity of deficit spending, Bush may be forced to address the need for higher taxes to pay for wartime outlays. It would be a black mark against Bush if he undid the efforts of his father and his predecessor by leaving behind red ink for future generations to face.

DON'T ATTACK YOUR OPPONENTS

No matter how heated politics got, Churchill and Roosevelt avoided divisive rhetoric. Where each might well have excoriated their political opponents for stubbornly refusing to recognize the looming danger of war, they always resisted pointing fingers or affixing blame, and instead embraced their opponents in a spirit of national unity. Johnson, in a different situation, was savage in attacking the domestic opponents of the Vietnam War. His war against his critics undermined his battle against the enemy and sapped any sense of national purpose, nobility, or unity. Bush has yet to take much fire from political critics. He must avoid party politics and fierce counterattacks, and focus on keeping the nation united.

DON'T FOOL YOURSELF

In the 1940s, the leaders of the United States and Britain grasped the dire realities they had to face. Neither deceived himself that the task was simple or that victory would come easily. Johnson was so intent on fooling America that he succeeded in misleading himself. As phony body counts of the enemy dead poured into the White House, he grasped at illusions to preserve his sense that victory was imminent. Bush seems to be avoiding self-deception, and he appears blessed with responsible advisors who keep their eyes open to facts and the truth. But only time will tell us if he maintains the firm grip on reality that characterized the leaders of World War II.

CHURCHILL AND ROOSEVELT MOBILIZE THEIR NATIONS FOR WAR, WHILE LYNDON JOHNSON MOBILIZES ONLY MISTRUST AND OPPOSITION

From the very first moments of World War II, Winston Churchill and Franklin D. Roosevelt made every effort to explain clearly and dramatically to their countrymen how dire an emergency they faced. Using the danger to catalyze their people to action, they called stirringly for sacrifice—and for a full commitment to victory.

Taking office in 1940 as French forces were being routed by the Germans, Churchill immediately set to work impressing upon Britons the uniqueness and magnitude of the task they faced, its historic importance, and the harsh consequences of failure. Seeking to challenge and energize rather than to reassure and soothe, Churchill used the truth as a weapon to mobilize his nation.

Churchill was quite clear in explaining the gravity of the moment. In the June 1940 speech "Their Finest Hour," he told his countrymen, "I do not at all underrate the severity of the ordeal which lies before us; but I believe our countrymen will show themselves capable of standing up to it . . . and will be able to stand up to it, and carry on in spite of it."

For Churchill, the moment was everything. With drama, courage, and conviction, he rallied his nation to face Hitler bravely. His passionate insistence that England would indeed win the war buoyed his countrymen during their worst times. And he summoned the British sense of history at the country's most desperate moment: "If the British Empire and its Commonwealth last for a thousand years, men will still say: this was their finest hour."

Across the ocean, President Franklin D. Roosevelt rallied America as all of its Pacific battleships were either destroyed or lay at the bottom of Pearl Harbor, permanently entombing hundreds of dying sailors. Calling the date of the attack, December 7, 1941, "a date which will live in infamy," he vowed that Japan's sneak attack would galvanize America, and that "long will we remember the character of the onslaught against us."

In a Fireside Chat the next day, Roosevelt was equally blunt in underscoring the defining challenge America faced in the twentieth century. "Every single man, woman and child is a partner in the most tremendous undertaking of our American history." Two months later he returned to the theme, telling the nation that "the task that we Americans now face will test us to the uttermost. Never before have we been called upon for such a prodigious effort. Never before have we had so little time in which to do so much."

Not only were the stakes greater than ever, Roosevelt warned, but the character of the conflict would be more difficult. "This war is a new kind of war. It is different from all other wars of the past, not only in its methods and weapons but also in its geography. It is warfare in terms of every continent, every island, every sea, every air-lane in the world."

Where Churchill and Roosevelt drew inspiration and energy from the gravity of the moment and boldly confronted their people with the magnitude of the task ahead, Lyndon Johnson avoided the issue in Vietnam. Rather than stir the nation to extraordinary effort, he slid into war almost imperceptibly. He sought to stress not the novelty of the crisis, but its routine resemblance to the other tests the nation had faced in its two decades of cold war. There was never a defining moment that opened the crisis—or even a declaration of war or a decisive congressional resolution to give the struggle a name or purpose. America moved toward war, as David Halberstam put it, "in such imperceptible degrees that neither the Congress nor the press could ever show a quantum jump. All the decisions were being cleverly hidden."

Obviously, the impetus for American involvement in the Vietnam War was very different from that which impelled our entry into World War II. There was no Pearl Harbor attack in the Vietnam War to wipe away any question that the United States would have to fight with all its power. But, lacking such a spur to action, Johnson never attempted to offer a patient, reasoned explanation of why Americans should feel compelled to fight and win in Vietnam. Doris Kearns Goodwin has described the inexorable and almost passive way Johnson entered the war in Vietnam: "Seated on the back of a beast that was far wilder than he had imagined, Johnson found himself carried along by its momentum, moving inexorably toward the wider war he did not want."

Johnson tried constantly to minimize the war and camouflage his policy. He thought that "it would be better not to frighten" Americans by telling them about the war, Halberstam notes, "or confront them too openly with it." Indeed, Johnson seems to have denied "even to himself the magnitude" of the problem in Vietnam, as historian Paul Conkin has written. Until midway through 1965, Johnson believed he could prevail in Vietnam "without major costs and without alarming the American public or inducing a warlike crisis . . . he wanted Vietnam to be a little noticed, backhanded operation." The result was that Johnson "did almost nothing to prepare Americans for a long and costly war."

While Johnson played down the seriousness of Vietnam, Roosevelt told the nation that "on the road ahead there lies hard work—grueling work—day and night, every hour and every minute." But FDR did not consider it a sacrifice to fight for freedom, he said. "I was about to add that ahead there lies sacrifice for all of us. But it is not correct to use that word. The United States does not consider it a sacrifice to do all one can, to give one's best to our nation, when the nation is fighting for its existence and its future life."

At the core of the difference between the Roosevelt-Churchill and Johnson approaches rests a basic disagreement about the willingness of their people to sustain a long and costly military commitment. FDR and Churchill deeply believed that their nations would rise to the challenge and be emboldened by learning of its gravity. Having seen Harry Truman's popularity fall during the Korean War, Johnson feared he could never convince the American people of the need to fight in Vietnam. "If there is one thing . . . that the American people will not take," he said, "it is another shooting war in Asia."

Worried about a public backlash, Johnson struggled to keep his plans to fight in Vietnam secret during his 1964 campaign against Barry Goldwater. Eager to paint the Republican as trigger-happy, he mouthed the antiwar sentiments he thought the nation embraced, declaring that "we are not about to send American boys nine or ten thousand miles to do what Asian boys ought to be doing for themselves."

But then, immediately after the election, Johnson secretly agreed to escalate the war. Even as he was sure that "those people out there don't want to go to war. They don't want a war in Vietnam," he made concrete plans to fight one.

The obvious question, as Kearns Goodwin asks, is: "How could Johnson have imagined that he could conduct a major war in virtual secrecy?"

Indeed, how did he think that he could mobilize public sentiment for the war without openly facing it and making his case to the people? If Johnson's secrecy was motivated by a belief that Americans didn't want to fight in Vietnam, didn't he realize that he had to persuade them to do so?

Instead, Johnson went to war secretly, manipulating the release of information to suit his purposes. Johnson thought, wrongly, that the American people didn't want to know what was happening. "Johnson's concept of the President's role in foreign policy reinforced his confidence that it would not be necessary to make full disclosure." As Kearns Goodwin writes, Johnson thought "the public . . . would only hurt itself by knowing too much."

So he lied.

By April 1, 1965—five months after he was elected as the "peace" candidate—Johnson had upped the ante in Vietnam, secretly committing more troops and fundamentally changing the character of the war. His press spokesmen believed their job was "to avoid clarifying the changes in the policy," Halberstam writes, "to misinform the public rather than inform it." For months they denied that anything had changed in the administration's war policy. Eventually, though, State Department briefing officer Robert McCloskey felt the need for more candor, and on June 7, 1965, told a press conference that the administration's policy had changed and U.S. forces were entering a ground war. When Johnson learned about the disclosure he "went into one of his wildest rages" and the White House sent out "the most vehement of denials" that anything had changed.

To the public, Johnson would admit only to sending 50,000 troops to Vietnam, not the 100,000 to 125,000 he had secretly ordered. He personally told one reporter that the increases in troop strength were just rumors and gossip, and that all he was doing was "filling out a few units." And he said that sending new troops to Vietnam "does not imply any change in policy whatever. It does not imply change of objective." As Halberstam writes, "In effect, the Administration was going to war without really coming to terms with it; they were paying for the war without announcing it or admitting it. Faking it." In the last analysis, LBJ simply didn't trust the American people to support him—and he was forced to deceive his people to hide the increasing toll the war would take on American money and lives.

While Churchill was quite open throughout the 1930s about his view that Hitler had to be stopped and that Britain needed to go to war to do it, FDR realized that his nation was not yet ready to enter a European war. Like Johnson in 1964, Roosevelt told campaign audiences as he ran for a third term in 1940, "I have said this before, but I shall say it again and again and again: Your boys are not going to be sent into any foreign wars."

But unlike Johnson, Roosevelt had also been laying the groundwork for American involvement in the war through a careful series of statements he had made throughout the late 1930s. In 1937 he warned of fascist expansionism, and called for a "quarantine" of the aggressor states. When war broke out in Europe he pledged American neutrality, but added that he was not asking Americans to be "neutral in thought as well," and made no secret of his sympathy for the Allied cause. As Britain faced its most trying moments, Roosevelt called for an end to the isolationist ban on aid to a belligerent—a risky decision two months before the 1940 elections—and gave Britain destroyers to keep sea-lanes open to the United States, swapping them for bases in the Caribbean and the Atlantic. In 1940, Roosevelt got Congress, by one vote, to pass legislation authorizing a military draft.

The ultimate step came in March 1941 when Roosevelt pushed through the Lend-Lease program of wartime aid to Britain.

Innocuously comparing the international situation to a community where a neighbor's house was on fire, FDR asked if it was not wise to lend him "my garden hose" to put the fire out before it spread to his own house.

Thus, long before the bombs fell on Pearl Harbor and catapulted the United States into war, FDR had carefully led the nation step by step up the ladder to full commitment. Where Johnson tried to conceal his escalation, Roosevelt had made every effort to condition Americans to the reality that war would soon be upon them. FDR carried his people along with him; Johnson left his in the dark.

Like most presidents before him, George W. Bush had spent little time trying to alert Americans to the danger of terrorism before the September 11, 2001, attacks on the World Trade Center and the Pentagon. Widely perceived as uncomfortable in the international sphere, he rarely spoke about terrorism, instead devoting the usual amount of attention to America's evolving relationship with Russia and China.

But as soon as the attacks hit, Bush recognized the need to tell his nation, frankly and openly, that a long struggle lay ahead. Calling the war on terror "the first battle in the war of the 21st century," Bush made clear the historic nature of the new challenge that confronted the nation. "I think the American people do understand," he said, "that after September 11th . . . we're facing a different world. And they accept that responsibility."

Immediately after the World Trade Center towers fell, Bush defined America's task not just as punishing those responsible, but as rooting out terrorism throughout the world. Unlike Clinton, who had treated the 1993 Trade Center bombing as a law enforcement problem, Bush pledged to hunt down each and every terrorist responsible. He defined the task as a military one, calling for war, and not just a police effort. He expanded the mandate of the struggle from simply catching those who had committed the outrage to punishing and eliminating the regimes that sponsored or protected them. Stressing that the war would take years, Bush was frank from the very beginning about the magnitude of the test we faced.

Sounding much like FDR, Bush told the press in October that "this is an unconventional war." It was a theme he would return to like a mantra. "It's not the kind of war that we're used to in America.

The Greatest Generation was used to storming beachheads. . . . Generation X was able to watch technology right in front of their TV screens. . . . This is a different kind of war that requires a different type of approach and a different type of mentality."

Bush even spoke of the war as a defining event in the evolution of America's national character, saying that Americans would feel "guided by a greater sense of purpose than any time during our lifetimes." Noting the often self-centered and hedonistic tenor of our times, he said, "some important things in our culture seem to be shifting" and he noted that we are "reminded of the true values of life."

How similar are Bush's statements to Churchill's that the "British nation is stirred and moved as it has never been at any time in its long, eventful, famous history." Or to Roosevelt's, when he declared that "this generation of Americans has come to realize, with a present and personal realization, that there is something larger and more important than the life of any individual or of any individual group— something for which a man will sacrifice, and gladly sacrifice, not only his pleasures, not only his goods, not only his associations with those he loves, but his life itself."

And, again, how different were the actions of Lyndon Johnson, who failed even to attempt to invest his people with a shared sense of mission. Vietnam had come as an interruption in his popular and public commitment to bring about a "Great Society" in America. Johnson's success on the civil rights and antipoverty fronts had given his administration hurtling momentum on the domestic scene. Not a man easily deterred from his chosen path, Johnson brushed aside the distraction of Vietnam by insisting that we could have our cake and eat it too. "This nation is mighty enough," he told Congress and the people in his 1966 State of the Union address, "to pursue our goals in the rest of the world while still building a Great Society here at home."

When Johnson's budget director, Charles Schultze, urged a $5 billion tax hike, Johnson's lack of faith in the American electorate led him to cut the increase to a more politically expedient $1 billion, mortgaging the nation's economic future rather than making a candid call for sacrifice in the present. In doing so, Johnson missed the opportunity to lead and inspire his nation. As Kearns Goodwin notes, "the business of war involves the severest sacrifices falling on the ordinary men and women in the country. Here, more than anywhere, the people must have an opportunity to make a choice." But Johnson denied Americans the chance to choose. He hid from them facts they could have used to decide either to make a serious national commitment—or to withdraw from the conflict. As Kearns Goodwin concludes, "in the end, no statesman can pursue a policy of war unless he knows for what goals, and for how long, his people are prepared to fight."

Johnson missed a basic point. Had he inspired the nation to meet a challenge larger than itself, he might have kindled the popular enthusiasm for the war effort that it required. As military historian Louis Heren noted, "most people associate war with some sort of sacrifice and austerity and [believe] that pain actually can build civilian morale by convincing people that they 'are doing their bit for the war effort.'" Johnson treated sacrifice as a political peril to be avoided, not as an incentive to catalyze commitment and energy.

Indeed, in his 1966 State of the Union address, Johnson seemed to deride the very idea of sacrifice. "There are men who cry out: We must sacrifice. Well, let us rather ask them: Who will they sacrifice? Are they going to sacrifice the children who seek the learning, or the sick who need medical care, or the families who dwell in squalor now brightened by the hope of home? Will they sacrifice opportunity for the distressed, the beauty of our land, the hope of our poor?"

By stressing that sacrifice was wrong—even morally evil—Johnson lent an aura of unreality to the war effort. Some should go abroad and die or be maimed for life, he seemed to say, but the rest of us must never be asked even to pay more in taxes or do without a social program. It was a disconnect that ate away at national morale.

By contrast, Roosevelt was clear in explaining the extreme challenges the war would impose upon the nation. Warning, two days after Pearl Harbor, that the "casualty lists of these first few days will undoubtedly be large," he spelled out in detail the kind of sacrifice he envisioned. "There will be a clear and definite shortage of metals for many kinds for civilian use, [which] we shall need for war purposes. . . . we shall have to give up many things entirely. And I am sure that the people . . . will cheerfully give up those material things that they are asked to give up."

Saying frankly that Americans must give up not only luxuries but "many other creature comforts," Roosevelt demanded higher taxes. He pushed through wage, price, and rent controls, and insisted on rationing "all essential commodities." Calling on Americans to meet the challenge in soaring terms, FDR told his radio audience that "the price for civilization must be paid in hard work and sorrow and blood. The price is not too high. If you doubt it, ask those millions who live today under the tyranny of Hitlerism.

"Never in the memory of man," he said, "has there been a war in which the courage, the endurance, and the loyalties of civilians played so vital a part."

Just as Churchill appealed to his nation's sense of history, Roosevelt invoked the first president's dark days at Valley Forge, saying, "Washington's conduct in those hard times has provided the model for all Americans ever since—a model of moral stamina."

Whereas Johnson would not cut back on the Great Society as he waded into Vietnam, FDR made it clear that the New Deal would have to be constrained as the war raged. The WPA and other New Deal programs were discontinued, as wartime priorities displaced Roosevelt's Depression-era services. At one press conference he memorably declared that "Dr. New Deal" was being replaced by "Dr. Win-the-War."

Which course will Bush take in the face of the inevitable economic challenge ahead? His trillion-dollar tax cut over the next ten years would seem as precious to him as the Great Society was to Johnson and the New Deal to Roosevelt. But as the costs of fighting the war on terrorism mount, consuming the budget surplus on which it is predicated, will Bush insist on leaving his tax cut untouched? Will he, like Johnson, pretend we can do it all and let deficit spending finance the war so that voters can still get refund checks in the mail? If our soldiers should begin dying in Afghanistan or Iraq, will he struggle to preserve both the nation's commitment to the war and his generous tax cuts?

History would suggest that Bush's willingness to give up his priorities, even as he asks us to give up ours, will be an important factor in his ability to sustain a national commitment to the war that transcends party and ideology. If he won't give up his tax cuts, why should the Democrats give up their spending programs? And why, indeed, as many asked in Vietnam, should some be asked to give up their lives?

As World War II progressed, both Roosevelt and Churchill understood that while they could not offer good news in the early years of the war, they must, at all costs, tell the truth. They realized that morale and national commitment hinged on their credibility in frankly and boldly facing and articulating the dangers that lay ahead.

From the moment of the attack on Pearl Harbor, Roosevelt acknowledged the magnitude of the disaster. As he faced Congress the day after the raid, he said somberly, "the attack yesterday on the Hawaiian Islands has caused severe damage to American naval and military forces. Very many American lives have been lost. In addition American ships have been reported torpedoed on the high seas between San Francisco and Honolulu. . . . There is no blinking the fact that our people, our territory and our interests are in grave danger."

In his Fireside Chat two days after Pearl Harbor, Roosevelt braced the nation for the trials ahead: "We must share together the bad news and the good news, the defeats and the victories—the changing fortunes of war. So far, the news has been all bad. We have suffered a serious setback in Hawaii. Our forces . . . are taking punishment. . . . The reports from Guam and Wake and Midway Islands are still confused, but we must be prepared for the announcement that all these three outposts have been seized."

Here, at the very outset of the war, he laid out the government's commitment to veracity. "This Government will put its trust in the stamina of the American people, and will give the facts to the public. . . . Of necessity there will be delays in officially confirming or denying reports of operations, but we will not hide facts from the country if we know the facts and if the enemy will not be aided by their disclosure."

Two weeks later, Churchill echoed FDR's call for truth in a speech to a joint session of the U.S. Congress: "Some people may be startled or momentarily depressed when, like your President, I speak of a long and hard war. But our peoples would rather know the truth, sombre though it be."

The contrast between their truthfulness and Johnson's deceit could not have been greater. Having begun the war by concealing America's entry into it, Johnson claimed that he could discern good news at every step of the way as he escalated our involvement. He assumed that Americans could not take bad news.

He filled the air with false and misleading optimism about the progress of the war. "Public pronouncements soon became a litany of optimistic estimates," Paul Conkin writes, "of endless progress reports that never seemed to cumulate in any significant improvement."

Using spurious body counts of enemy dead, Johnson kept telling America that we were winning the war. "The enemy is no longer close to victory," he enthusiastically told Congress in 1966. "Time is no longer on his side. There is no cause to doubt the American commitment."

Yet even as he spoke the American commitment was eroding, along with LBJ's credibility. When Americans began to add up the numbers of supposed North Vietnamese and Viet Cong dead the administration claimed to have killed in its daily barrage of optimism, the totals became so ridiculously large as to defy reason. "By late 1966, the credibility gap was eating away at the administration," Joseph Califano, Jr., wrote in a 1991 biography. In a working meeting on the 1967 state of the union message, Johnson aide Harry McPherson complained, "The President is simply not believed."

Of course, the larger-than-life Johnson had always been known for exaggeration. When he implied that he was a latter day Lincoln who had grown up in a rural shack, his mother corrected him: "Why, Lyndon, you know that's not true, you know you were born and brought up in a perfectly nice house much closer to town."

But if his exaggerations about progress in Vietnam "seemed amusing early in the game," Halberstam writes, "later, as the pressure of Vietnam mounted and the President's credibility problems centered on greater issues, they would not seem so amusing."

Johnson's paranoia about secrecy, and his refusal to lay the facts before the public, drained the nation's morale, vitality, and self-confidence, as Americans gradually came to realize that their own government was lying to them. Johnson's web of deceit embraced his White House staff, his cabinet, and even his generals. When one aide warned the president that the U.S. commander in Vietnam, General

William Westmoreland, "might go public with the war," Johnson reacted sharply: "No, he won't . . . because I'm the one person who has what he wants." By that, Halberstam infers, LBJ meant a "future promotion."

Not only did Churchill and Roosevelt refuse to pass on (or manufacture) false good news, they did their best to temper optimism lest it lead to overconfidence. "I deprecate premature rejoicings," Churchill told the British people on September 30, 1941, "and I indulge in no sanguine predictions about the future. . . . We must not, I repeat, relax for an instant."

When the British saved the bulk of their army, and a good part of France's, in the miraculous evacuation from Dunkirk, he warned his celebrating and grateful nation that "wars are not won by evacuations." Even when he had good news to report, Churchill warned against optimism. "We must be on our guard equally against pessimism and against optimism. . . . [Events] must not lead us for a moment to suppose that the worst is over." Even after the success of the Battle of Britain between the Royal Air Force and the Luftwaffe, Churchill noted that "the mood of Britain is wisely and rightly averse from every form of shallow or premature exultation. This is no time for boasts or glowing prophecies."

When the war in Europe was over and Britain and American arms had triumphed, Churchill still refused to yield to optimism, warning about the need to win the war in the Pacific and to face the menace of communism in Europe: "I wish I could tell you tonight that all our toils and trouble were over. Then indeed I could end my five years' service happily. . . . But, on the contrary, I must warn you, as I did when I began . . . that there is still a lot to do, and that you must be prepared for further efforts of mind and body and further sacrifices. . . . You must not weaken in any way."

Churchill and Roosevelt were both candid in saying that there would be occasions when they would not lay out the entire truth to their peoples. This very admission made a certain necessary level of deception palatable to the public in both lands.

Churchill warned the House of Commons that secrecy would be an important weapon in the war. "The House will remember that in June last I deprecated the making of too frequent expositions of Government policy and reviews of the war situation by Ministers of the Crown. Anything that is said which is novel or pregnant will, of course, be studied attentively by the enemy and may be a help to him in measuring our affairs. . . . For reasons which I have explained . . . we have since June abandoned the practice of publishing statements at regular monthly intervals of our shipping losses, and I propose to continue this salutary practice." He told the nation that, "If I were to throw out dark hints of some great design, no one would have any advantage but the enemy."

In the first months of the war against terrorism, George W. Bush has also avoided the pitfalls of false optimism. He has taken great pains to warn America of the challenges abroad and the dangers at home. As the *New York Times* noted: "Mr. Bush tried to prepare the country for a long and potentially costly war. There will be no easy victories, despite the early success of American air strikes in Afghanistan. . . . Mr. Bush was effective in talking to the American people about their fears. He spoke candidly about new warnings that additional terrorist attacks could come at any time."

Bush has also been clear in explaining the importance of secrecy to the conduct of the war. Warning Americans in his speech to Congress that the war on terrorism will include "covert operations secret even in success." Bush has established a legitimate zone of secrecy to avoid falling into the Johnsonian pitfalls of deception.

But as the course of the war is marked by transitory successes and momentary victories, Bush must avoid scuttling his credibility on the rocks of premature optimism. Nothing so energizes a people as challenge—and nothing so enervates them as false confidence and premature celebration.

The deception for which Johnson paid the highest political price was that which he practiced on Congress. Where Churchill and Roosevelt got formal declarations of war from Parliament and Congress, and Bush got a joint resolution authorizing military action, Johnson predicated the war in Vietnam on the shaky Gulf of Tonkin Resolution of 1964.

The full facts of the naval engagement between American and North Vietnamese ships that triggered the resolution may never be fully determined. But what is clear is that the powers Congress voted to the president in the resolution went far beyond what they intended at the time. As Conkin notes, the congressmen and senators who supported the Gulf of Tonkin Resolution "did not fully understand the complexity of the problems in South Vietnam, did not embrace a clear set of goals, and did not foresee the enormous cost ahead. But neither did the Johnson administration."

Exploiting the open-ended authorization granted the president in the resolution, Johnson went to war without taking Congress fully into his confidence. Indeed, the war in Vietnam only really came to an end a decade later, when Congress voted to curtail appropriations for a war on which it had never really been consulted.

When Churchill took office, he immediately set about to include the opposition Labour Party in his government. Upon becoming prime minister in May 1940, the historian Martin Gilbert writes, Churchill "sent a message to Clement Attlee [the Labour Party leader], asking to see him. Attlee came, together with [Labour Party MP] Arthur Greenwood, and when Churchill asked if the Labour Party would join his Government, Attlee gave his assent. . . . Churchill asked Attlee to let him have a list of Labour men to consider for particular offices." Fashioning a wartime unity government, Churchill made Attlee deputy prime minister, and included a large number of Labour Party leaders in his government.

Churchill could have exacted retribution from his political foes, like former prime minister Neville Chamberlain, who had kept him out of office for years while he warned from the sidelines about the growing menace of Nazi Germany. FDR must have similarly been tempted to rub the Republican isolationists' noses in the dirt after their years of refusal to arm America had weakened the nation. After all, their opposition had almost killed Roosevelt's efforts to begin a military draft. Both men would have had a good case had they sought revenge. But neither man gave into the temptation.

To the contrary, Churchill cheerfully included his predecessor in his cabinet, refusing to "succumb to the cry of those who urged the immediate ousting of Chamberlain. . . . A year earlier he had been the severest of [their] critics. Now he rejected all pressures for a cutting off of heads," Gilbert writes.

Roosevelt was similarly committed to bipartisanship as the war approached. He decided to create a new, unity war cabinet, a move that would insulate him from partisan bickering as he pursued the rearming of the nation. FDR named Republicans Henry Stimson as his secretary of war and Frank Knox as secretary of the navy. As he reached out to the other party, historian Jeffrey Hacker writes, "Congress was solidly behind him . . . indeed, the war made Roosevelt more than a political power broker, more than a military commander. He became the standard-bearer for all Americans."

After Pearl Harbor, patriotic fervor immediately overwhelmed partisan rancor. A January 1942 poll confirmed that 84 percent of Americans approved of Roosevelt's leadership. A similar spirit, of course, gripped America in the wake of the September 11 attacks, and George W. Bush's approval ratings have soared even higher—by some counts well into the nineties. And so far Bush has admirably followed in Churchill's and Roosevelt's footsteps in his almost daily consultations with Democratic legislative leaders Senator Tom Daschle and Congressman Richard Gephardt. "They are his new best friends," one top Republican political operative told me in October 2001. "He takes them with him everywhere he goes and seems to talk to them every hour."

When the going gets rough, Bush, like Churchill and Roosevelt, seems mindful that he will need congressional and Democratic support for the war effort. His decision to have the opposition on his side is crucial to sustaining the war.

Johnson, on the other hand, created and exacerbated an air of political rancor that won him plenty of enemies throughout his pursuit of the Vietnam War. Using all the tools of the presidency, he attacked, ridiculed, and derided critics of the war, impugning their patriotism and darkly implying that they were selling out American troops in the field. His attacks helped to polarize America, planting the seeds of an "us-versus-them" mentality that pitted doves against hawks. "Wars are supposed to unite nations, to rally divided spirits," David Halberstam observes, "but this war was different; rather than concealing or healing normal divisions in the society, it widened them, and gaps became chasms."

But Johnson's attempts to fool the American people, the media, and Congress pale by comparison with his effort to fool himself. The roots of Lyndon Johnson's ultimate failure lay in his success at self-deception. Johnson refused to hear anything but good news. Doubters, naysayers, critics, and skeptics were kept away. "It was very hard to bring doubts and reality to Johnson without losing access. The reasonable had become unreasonable; the rational, irrational. . . . The more outcry in the country . . . the more Johnson hunkered down, isolated himself from reality. What had begun as a credibility gap became something far more perilous, a reality gap."

According to Halberstam, key Johnson aide McGeorge Bundy recalled that there was a "premium put on imprecision." "If the military and political leaders had been totally candid with each other in 1965 about the length and cost of the war," Halberstam noted, the outcome might have been much different. Instead the administration pursued "a deliberate policy not to surface with real figures and real estimates which might show that they were headed toward a real war. The men around Johnson served him poorly, but they served him poorly because he wanted them to."

As Johnson's critics became his enemies, America found itself divided in war as it had not been since the Civil War. Mocking his adversaries, withholding information, misleading the media, Lyndon Johnson presided over a nation divided into two camps—those who went along with his self-delusion and those who did not.

Churchill, on the other hand, boldly confronted reality, as his gritty remembrance of the days of the Luftwaffe blitzkrieg over London reveals. "Our outlook at this time was that London, except for its strong, modern buildings, would be gradually and soon reduced to a rubble-heap. I was deeply anxious about the life of the people of London." Facing disaster as France fell, Churchill's aide General Sir Edmund Ironside wrote in his diary on June 17, 1940, of his leader's resolution in facing reality. "There is no doubt that Winston has any amount of courage and experience. Thrown with his back to the wall, he may lose some of his lack of balance . . . [but] he is quite undismayed by the state of affairs." Ironside's replacement, General Alan Brooke, wrote in his diary that he found Churchill "full of the most marvelous courage considering the burden he is bearing. . . . I think he fully realises the difficulties he is up against."

Will Bush maintain his sense of perspective as he prosecutes a war against an essentially unknowable enemy? The view from 1600 Pennsylvania Avenue can often be thick with fog. Fog, of course, is caused by a clash of warm and cold air. When the hothouse of comforting

White House staff adulation meets the frost of its harsh critics outside, a fog can develop that obscures the president's vision and beclouds his sense of reality.

The key to maintaining a realistic view and avoiding self-deception is to bring the critics inside with their arguments and to go outside to make the case to the country. By tempering the blind loyalty inside the building with other views and contesting the criticism outside it by offering facts, a strong president can keep White House windows clear and free of mist throughout even the gravest crisis—just as they were in the 1940s.

EPILOGUE

The strategies the politicians in this book have used not only proved to be effective ways to achieve power but tended to leave their democracies better for the experience. As each figure pursued his own self-interest, he ended up by advancing his country's political system at the same time.

The appeals to patriotism Reagan, De Gaulle, Churchill, and Lincoln used to vindicate their lifelong convictions obviously brought about tremendous progress. The great deeds they each did would likely not have been possible had they failed—as Wilson, Gore, and Goldwater failed—to expand their parochial message to a nationalist agenda.

By triangulating, George W. Bush and Bill Clinton each moved his political party ahead and forced it to embrace new solutions to problems that had proved intractable to their predecessors. When Bush made the Republicans include a federal involvement in education, and Clinton forced his party to wind down the vicious cycle of welfare dependency, each won an election—but also made important substantive gains on crucial national issues. When Mitterrand abandoned his party's historic commitment to government ownership of industry, he began the process of transforming his Socialist Party into a party of the Left that could thrive in a neoliberal world.

When Abraham Lincoln divided the Democrats and conquered them, he opened the way for the eradication of the great national evil of slavery. With only 40 percent of the vote in the 1860 election, it is hard to see how else he could have achieved the mandate to change our nation so totally. Yet his divide-and-conquer approach stands in stark contrast to Richard Nixon's, which—like much else in his record—remains without redeeming virtue.

Blair's and Koizumi's party-reform movements moved their nations' political process further down the path toward democracy, breaking the power of the unions in one country and the political elites in the other—while George McGovern's example bears witness to the harsh self-sacrifice that can await the reformer at the end of his political road.

The Fireside Chats of FDR and the presidential debates of Kennedy were breakthroughs in public participation in the democratic process. Both raised information levels to new heights and gave the average voter a way to second-guess the intermediaries who used to barter their votes for special favor. Was Lyndon Johnson's invention of the negative ad a positive achievement for our democracy? You bet it was! Harsh though the results may sometimes be, voters are now able to use the competition of the political marketplace to hold their elected officials responsible for sins that once were swept under their office rugs.

And when Roosevelt, Churchill, and—so far—George W. Bush told their citizens the truth about the challenge they faced and the sacrifices it necessitated, they encouraged national unity and leavened their lands with a maturity and a dedication that helped to transform their people—as Lyndon Johnson had failed to do in the 1960s.

Will the new techniques and technologies of the future improve our political process? Likely they will. Each new development seems to move in the same direction—toward greater direct democracy and more public participation. Certainly the Internet and the two-way political dialogue it fosters will tend to make our democracy more direct and citizen involvement more consuming.

But it is, perhaps, the almost offhand way in which politicians do good while they are attempting only to do well that is the ultimate lesson of this volume. Even as they were pursuing their own personal ends and gratifying their instinct for the acquisition of power, in countless instances like these, politicians have helped us all and catalyzed national progress. It is this lesson, as counterintuitive as it is, that I find most compelling in surveying the course of political history: that the process itself tends to serve best the men who trust—and serve—the people the most.

ACKNOWLEDGMENTS

First and foremost, I would like to deeply thank my wife, Eileen McGann, for her ideas and creative contributions to this book. But I am especially grateful to her for imparting order, logic, and organization to the editing, checking, and verifying of so many facts, quotes, and figures. She was a lifesaver.

I would also like to thank the small army of talented family and friends she assembled to pore over the data and smooth the rough edges. If Eileen was the commander, Mary McGann, her sister, was her right hand as they pored through the copy. This tireless and dedicated group included Debbie Chang, Joel Morton, Katie Maxwell, Maureen Maxwell, Elizabeth Jacoby, and Thomas Gallagher.

Leslie D. Feldman, on the faculty at Hofstra, assembled a lot of the basic research information and organized it to make it userfriendly. She helped to shape the argument in many chapters.

Marco Desena, an extremely talented and energetic student at Baruch, hustled around New York getting books, copying articles, putting together the material, extracting information, and developing the data.

Zach Goldfarb, a brilliant student at Princeton, did the excellent basic research on three of the chapters: Blair, Koizumi, and Roosevelt.

Marco and Zach have illustrious futures ahead of them.

I would also like to thank the production, art, marketing, and publicity staff of ReganBooks and HarperCollins, including Lina

Perl, Andrea Molitor, Cassie Jones, Dan Taylor, Kurt Andrews, Carl Raymond, and Jennifer Suitor.

In a way, I owe some of the inspiration for this book to my mates at Fox, Bill O'Reilly and Sean Hannity, who got me thinking. Bill and Sean would always listen to my discussions of political maneuvers on their shows and ask, "What about standing on principle?" My answer became the first chapter of this book.

Calvert Morgan of ReganBooks is an editor who shapes the work with uniquely prescient guidance and input.

Judith Regan, my publisher, is a publishing genius. We all know that. But she's also a creative force of great power and vision. She's the one who suggested this book in the first place, conceived its title, and sketched out its outline. Wow.

Notes

I. STAND ON PRINCIPLE

Ronald Reagan

Reagan's own speeches specifically consulted include: commencement address at his alma mater, Eureka College, on June 7, 1957; "A Time for Choosing," his speech introducing Barry Goldwater during 1964 presidential campaign (October 27, 1964); several speeches to the Conservative Political Action conferences ("Shining City on a Hill," January 25, 1974; "Let Them Go Their Way," March 1, 1975; "The New Republican Party," February 6, 1977; "America's Purpose in the World," March 17, 1978); his acceptance of the Republican nomination for President, July 17, 1980; the inaugural addresses (January 20, 1981, and January 21, 1985); his "Evil Empire" speech (June 8, 1982); and his farewell speech (January 11, 1989). Books by Reagan include his autobiography, *An American Life* (New York: Simon and Schuster, 1990), as well as *Where's the Rest of Me?* by Reagan and Richard Hubler (New York: Karz Publishers, 1965), and Abortion and the Conscience of the Nation (New York: Thomas Nelson, 1984). Secondary work on Ronald Reagan includes Lou Cannon, *President Reagan: The Role of a Lifetime* (New York: Simon and Schuster); *Dutch: A Memoir of Ronald Reagan* (New York: Random House, 1999), Reagan's official biography by Pulitzer Prize–winning biographer Edmund Morris; and Dinesh D'Souza, *Ronald Reagan: How an Ordinary Man Became an Extraordinary Leader* (New York: The Free Press, 1997).

page
1   "They did not see...": Ralph Waldo Emerson, "The American Scholar," speech to the Phi Beta Kappa society, Cambridge, Massachusetts, August 31, 1837.
2   "Persecution is an ... The real advantage ...": John Stuart Mill, "Of Thought and Discusson," in *On Liberty* (Indianapolis, Indiana: Bobbs Merrill, 1956).
3   "dustbin of history...": Leon Trotsky, *History of the Russian Revolution*, Vol. 3, Ch. 10 (1973). "How can you be...": quoted in *Newsweek*, October 1, 1962, p. 34.

8   "[The spectre] our well-meaning . . . we are being . . .": *Where's the Rest of Me?*,
      pp. 311–12; "This irreconcilable conflict . . .": commencement address, Eureka
      College, June 7, 1957; "The time has come . . .": "California and the Problem
      of Government Growth," governor's inaugural address, January 5, 1967.

9   "What we needed . . .": quoted in Goldwater, *With No Apologies* (New York:
      William Morrow, 1979), pp. 275–6; "I don't know about . . .": "Let Them Go
      Their Way," speech to the second Conservative Political Action Conference,
      March 1, 1975; "always been puzzled . . .": "The New Republican Party," speech
      to the fourth Conservative Political Action Conference, February 6, 1977; "a
      political party . . .": "Let Them Go Their Way," March 1, 1975.

10   "a banner of . . .": "Let Them Go Their Way," March 1, 1975; "Call it mysti-
      cism . . .": "To Restore America," speech, March 31, 1976.

11   "Planned parenthood is my baby": quoted in Lee Edwards, *Goldwater: The Man
      Who Made a Revolution* (Washington, D.C.: Regnery, 1995), p. 420; "I think
      every . . .": "Conservative Pioneer Became an Outcast," *The Arizona Republic*,
      May 31, 1998; "abortion on demand": quoted in *Goldwater*, p. 421; "combine
      the two . . .": speech to the fourth Conservative Political Action Conference,
      February 6, 1977.

13   "recognizes that national . . .": Burton Yale Pines, "The Ten Legacies of Ronald
      Reagan", speech at People's University of China, October 1988.

14   "We bought it . . .": quoted in Adam Clymer, "Mirror on the Past: Canal's Fate
      Reflects Shift by U.S.," *New York Times*, December 15, 1999; "though I am
      wounded . . .": quoted in ibid., p. 79.

15   "He recommended that . . . He suggested that . . .": ibid., p. 76; "It was a dread-
      ful . . .": "The Ten Legacies of Ronald Reagan."

16   "has no rudder . . . Is the world . . .": nomination acceptance speech, Republi-
      can National Convention, July 17, 1980; "Recession is when . . .": "Out of the
      Past, Fresh Choices for the Future," *Time*, January 5, 1981; "a radical depar-
      ture . . .": second presidential debate, October 28, 1980; "There you go again":
      quoted in D'Souza, *Ronald Reagan*, p 83.

Barry Goldwater

The primary sources on Goldwater include his own books: *The Conscience of a Conservative* (Louisville, Kentucky: Victor Publishing Co, 1960); *Where I Stand* (New
York: McGraw Hill, 1964); and his autobiography, *With No Apologies* (New York:
William Morrow, 1979). Secondary sources included Theodore White, *The Making
of the President 1964* (New York: Atheneum, 1965); Lee Edwards, *Goldwater: The Man
Who Made a Revolution* (Washington, D.C.: Regnery, 1995); Lionel Lokos, *Hysteria
1964: The Fear Campaign Against Barry Goldwater* (New Rochelle, New York: Arling-
ton House, 1967); John Kessel, *The Goldwater Coalition: Republican Strategies in 1964*
(New York: Bobbs Merrill, 1968); and Stephen Shadegg, *What Happened to Goldwater? The Inside Story of the 1964 Republican Campaign* (New York: Holt, Rinehart,
1965).

page
18   "I am quite certain . . .": *Where I Stand*, pp. 32–3; "the government must
      begin . . .": *The Conscience of a Conservative*, p. 66.

19   "The need [is] . . .": *Where I Stand*, p. 108; "The effect of Welfarism . . .": *The Conscience of a Conservative*, pp. 70–71; "would be better . . .": *The Making of the President 1964*, p. 104; "dime-store New Deal": quoted in ibid.

20   "the federal Constitution . . .": *The Conscience of a Conservative*, pp. 33–4.

21   "If a person . . .": Lee Edwards, "The Unforgettable Candidate," *National Review*, July 6, 1998; "Goldwater Sets Goals . . .": *Goldwater: The Man Who Made a Revolution*, p. 207; "My opponents built . . .": reported in *U.S. News & World Report*, December 21, 1964, and quoted in *Hysteria 1964: The Fear Campaign Against Barry Goldwater*.

22   "sometimes intemperate personality": *New York Times*, July 16, 1964, p. 17; "if either . . .": reported in *San Francisco Chronicle*, April 1, 1964, p. 1, and quoted in Lokos, *Hysteria 1964: The Fear Campaign Against Barry Goldwater*, p. 108; "to restore the strength . . .": quoted in *New York Times*, July 16, 1964, p. 17; "Now it would . . .": *The Conscience of a Conservative*, p. 63; "the biggest faker . . . phoniest individual . . .": *New York Times*, July 16, 1964, p. 1; "dangerous" and "frightening": quoted in *Goldwater: The Man Who Made a Revolution*, p. 276; "searching his conscience": quoted in ibid., p. 266; "giving the right . . ." quoted in ibid., p. 276.

23   "a party ready . . . to unify the country . . .": Reagan nomination acceptance speech, Republican National Convention, July 17, 1980; "elevate the state . . . resist concentrations . . .": Goldwater nomination acceptance speech, Republican National Convention, 1964; "We see in . . . decentralized power . . . Now, we Republicans . . . Anyone who joins . . .": ibid.

24   "extremism in the . . .": ibid.; "Reagan and Goldwater . . .": Michael Gerson, "Mr. Right," *U.S. News & World Report*, June 8, 1998; "crazy quilt . . .": *With No Apologies*, p. 186; "too casually prescribed nuclear . . .": quoted in *Hysteria 1964: The Fear Campaign Against Barry Goldwater*, p. 124; "negative image . . . so-called 'nut' . . . an undecided voter . . .": George H. W. Bush, "The Republican Party and the Conservative Movement," *National Review*, December 1, 1964.

25   "The whole campaign against me . . .": reported in *U.S. News & World Report*, December 21, 1964, and quoted in *Hysteria 1964: The Fear Campaign Against Barry Goldwater*, p. 13.

Winston Churchill

There are two indispensable sources on Winston S. Churchill: Churchill's own multi-volume history *The Second World War*, and the official biography by Martin Gilbert. Churchill's series includes *The Gathering Storm, Their Finest Hour*, and *The Grand Alliance* (Boston: Houghton Mifflin, 1948, 1949, 1950). Other primary sources include several collections of Churchill's speeches: *The End of the Beginning* (Boston: Little Brown, 1943); *The Unrelenting Struggle*, compiled by Charles Eade (Boston: Little, Brown, 1942); *Secret Session Speeches* (New York: Simon and Schuster, 1946); and *Blood, Toil, Tears and Sweat: The Speeches of Winston Churchill*, edited by David Cannadine (Boston: Houghton Mifflin, 1989).

The secondary literature on Churchill includes Martin Gilbert's official multi-volume biography; volumes consulted here include *Winston S. Churchill: The Prophet of Truth, Volume V 1922–1939* (London: Heinemann, 1976), and *Winston S. Churchill: Finest Hour, Volume VI 1939–1941* (London: Heinemann, 1983). His other volumes on Churchill include *Winston Churchill: The Wilderness Years* (Boston: Houghton Mifflin, 1982), and the single-volume *Churchill* (New York: Henry Holt and Co, 1991). Also of use were William Manchester's *The Last Lion*, particularly the second of its two volumes, *Winston Spencer Churchill: Alone*, 1932–1940 (Boston: Little, Brown, 1988); and J. R. Clynes, *Memoirs* (London: Hutchison, 1937).

Churchill's speeches used here include: "I Have Done My Best" (Dundee, June 5, 1915); his first speech as prime minister (May 13, 1940); "We Shall Fight on the Beaches" (June 4, 1940); "Their Finest Hour" (June 18, 1940); "We Are Vulnerable" (February 7, 1934); "The Outlook for 1942" (March 26, 1942); "Keep Right on to the End" (Edinburgh, October 12, 1942); and "We See the Ridge Ahead" (Leeds, May 16, 1942), as well as others. Most of the speeches are available on the Web.

page

27   "I have had the blame . . . I have done my best": June 15, 1915, quoted from *Blood, Toil, Tears and Sweat*, p. 61; "he would die . . .": quoted in Mary Soames, *Clementine Churchill: The Biography of a Marriage* (Boston, Houghton Mifflin, 1979); "without an office . . .": quoted in Gilbert, *Churchill*, p. 454.

28   "He was a delicious and . . .": Kingsley Martin, *Father Figures*, as quoted on *http://www.spartacus.schoolnet.co.uk/2WWChurchhill.htm*; "In dealing with Oriental . . .": "A Seditious Middle Temple Lawyer," Winchester House, February 23, 1931, from *Blood, Toil, Tears and Sweat*, p. 101.

29   "to all the world . . .": ibid, p. 101; "expensive toy . . .": quoted in *The Last Lion*, p. 557; "blocked everything on . . .": ibid., p. 218; "think . . . the English . . .": quoted in ibid., p. 636.

30   "One would imagine . . .": quoted in ibid., p. 652; "peace in our . . .": quoted from *Blood, Toil, Tears and Sweat*, p. 129; "In the five days . . .": *The Last Lion*, p. 1000; "total and unmitigated defeat . . .": "A Total and Unmitigated Defeat," October 5, 1938, from *Blood, Toil, Tears and Sweat*, p. 130.

31   "The partition . . .": press statement of September 21, 1938, quoted in *Winston S. Churchill: The Prophet of Truth*, p. 978; "I understood the Prime Minister's . . .": *The Gathering Storm*, p. 356; "thousands of enormous posters . . .": *The Gathering Storm*, p. 358.

32   "First Lord's Prayers" anecdote from Churchill Center Web page, autumn/fourth quarter 1939, *www.winstonchurch.org/action.htm*; "The Prime Minister . . .": from diary of Harold Nicolson, September 1939, quoted in *The Last Lion*, p. 601; "The effect of . . .": quoted in ibid., p. 602. "Oh, you can't do that . . .": quoted in ibid., p. 583.

33   "The idea of not . . .": quoted in ibid., p. 585.

34   "sprang out . . .": quoted in ibid., p. 561; "We tried again . . .": quoted in ibid., p. 601; "Now we have begun . . .": quoted in ibid., p. 601.

35   "the sensation and . . .": quoted in ibid., p. 595; "His natural aggression . . .": quoted in ibid., p. 574. "certainly gives one . . ." quoted in ibid., p. 594; "the views of Churchill . . .": quoted in ibid., p. 600; "So that is what . . ." quoted in ibid., p. 594.

36   "eleven inch guns . . . fifteen miles . . . sent nine British Cargo ships . . . gaping holes . . . limped . . . gave intense joy . . .": quoted in ibid., pp. 566–8.

37   "on the night of . . .": *The Gathering Storm*, pp. 666–7; "that this House . . .": address to House of Commons (first speech as Prime Minister), May 13, 1940.

40   "we shall not flag or fail . . .": "We Shall Fight on the Beaches," address to House of Commons, June 4, 1940; "if we fail, then the whole world . . .": "Their Finest Hour," June 18, 1940, quoted from *Blood, Toil, Tears and Sweat*, p. 34. "O Ship of State" by Henry Wadsworth Longfellow, available in *The Poems of Henry Wadsworth Longfellow* (New York: Modern Library, 1965), and many other editions.

Charles de Gaulle

The primary sources on de Gaulle are his three-volume set of World War II memoirs; *The Complete War Memoirs of Charles de Gaulle* (New York: Carroll & Graf, 1998), referred to below simply as *War Memoirs*, incorporates all three. Biographies include Jean Lacouture, *De Gaulle: The Ruler 1945–1970* (New York: W. W. Norton, 1991), and Don Cook, *Charles de Gaulle: A Biography*, (New York: G. P. Putnam's Sons, 1983); other secondary sources include William L. Shirer, *The Collapse of the Third Republic* (New York: Simon & Schuster, 1969), and Henry Kissinger, *Years of Upheaval* (New York: Little, Brown & Co., 1982).

page

44   "Deliberation is the work . . .": *War Memoirs*, p. 483; "arm in arm . . .": quoted in *The Collapse of the Third Republic*, p. 748.

45   "the weak gain strength . . .": *Years of Upheaval*, p. 173; "As long as . . .": *War Memoirs*, p. 497; "You will remain the only choice . . .": quoted in ibid., p. 362; "no longer inspired by principles . . .": ibid., p. 940.

46   "could result only . . . If the government fell into . . .": ibid.; "discovering with a . . .": quoted in ibid., p. 148; "As I saw it . . .": ibid., p. 273; "As the champion . . .": ibid., p. 271; "cooking its own . . .": quoted in *De Gaulle: The Ruler*, p. 143; "designated by the people . . .": *War Memoirs*, p. 941.

47   "dominance of . . .": *Charles de Gaulle*, p. 289; "now rapidly . . .": ibid.; "I could not overlook . . .": *War Memoirs*, p. 941; "political activity . . .": *Charles de Gaulle*, p. 288.

48   "Before a week . . .": quoted in *De Gaulle: The Ruler*, p. 124; "The exclusive regime of the parties": quoted in *Charles de Gaulle*, pp. 294–5, and in *War Memoirs*, p. 993; "with that . . .": ibid., p. 295; "Really, one can scarcely . . .": quoted in ibid., p. 294; "I have made at least one political . . .": quoted in *De Gaulle: The Ruler*, p. 124.

49   "to withdraw from events . . .": *War Memoirs*, p. 977; "I prefer my legend . . .": quoted in *De Gaulle: The Ruler*, p. 102; "to promote, above the parties . . .": quoted in Anthony Hartley, *Gaullism* (New York: Outerbridge and Dienstfrey, 1971), p. 136; "every group, party . . .": Andre Malraux and James Burnham, *The Case for de Gaulle* (New York: Random House, 1948), p. 9; "The efforts that . . .": quoted in *De Gaulle: The Ruler*, p. 153.

50   "we must not . . .": ibid., p. 105. "It was during . . . He acquired . . .": quoted in ibid., p. 153.

51   "The day will come . . .": quoted in ibid., pp. 135–6.

52   "This indeed was . . .": *Charles de Gaulle*, p. 319; "The degradation of the state . . .": quoted in *De Gaulle: The Ruler*, p. 169; "On May 19 . . .": *Charles de Gaulle*, p. 319.

53   "specify at the present time . . . I know of no judge . . .": quoted in ibid., p. 320;

"Now I shall . . .": quoted in ibid., p. 321; "the unity and independence . . .": quoted in ibid., p. 321; "Any action endangering . . .": quoted in ibid.; "I have called . . .": quoted in ibid., p. 322; "The Gaullist strategy . . .": quoted in *De Gaulle: The Ruler*, p. 176.

54   "All liberties . . .": quoted in *Charles de Gaulle*, p. 323; "govern by decree . . .": ibid., p. 323; "submitted for approval . . .": ibid; "Most Communists . . .": ibid., p. 324; "liberation all over . . .": ibid; "I have understood you . . .": quoted in ibid., p. 325; select and dismiss: ibid., p. 326.

55   "almost monolithic political power . . .": ibid., p. 327.

Abraham Lincoln

Primary sources include the speeches of Abraham Lincoln, texts of the Lincoln-Douglas debates (available in various edited collections), and Horace Greeley's exchange of public letters with Lincoln. Biographies and secondary sources include David Herbert Donald, *Lincoln* (New York: Simon and Schuster, 1995); Peter Burchard, *Lincoln and Slavery* (New York: Atheneum, 1999); and Eric Foner, *Free Soil, Free Labor, Free Men* (New York: Oxford University Press, 1970).

page
57   "was considered . . . to be . . .": *Free Soil, Free Labor, Free Men*, p. 214; "I have always hated . . . hatred to the. . . .": quoted in ibid., p. 215.

58   "the barbaric institution . . . ultimate extinction": quoted in ibid., p 215; "a house divided . . .": speech at Springfield, Illinois, June 16, 1858; "one of the major reasons . . .": *Free Soil, Free Labor, Free Men*, p. 219.

59   "Lincoln agreed . . .": ibid., p. 225; "The Republican position . . .": ibid., p. 225; "George Washington . . .": speech at Cooper Union, New York City, February 27, 1860.

60   "no purpose . . .": first inaugural address, March 4, 1861.

61   "Though passion may . . .": ibid.

62   "If I could save . . .": letter to Horace Greeley, published in the *New York Tribune*, August 22, 1862.

Woodrow Wilson

Primary sources are the public papers of Woodrow Wilson and his speeches—notably the League of Nations speech of September 25, 1919. A volume used here is Ray S. Baker and William E. Dodd, editors, *The Public Papers of Woodrow Wilson, Volume 1* (New York: 1924). Secondary sources include Robert Ferrell *Woodrow Wilson and World War I, 1917–1921*; Elmer Bendiner, *A Time for Angels* (New York: Random House, 1975); and Gene Smith, *When the Cheering Stopped* (London: Hutchinson and Co., 1964). Wilson's speeches can be found at *http://www.tamu.edu/scom/pres/ speeches/wwleague.html.*

page
66   "the world must be . . .": war message to Congress, April 2, 1917; "For many a long day . . .": State of the Union address, December 2, 1918; "peace without

victory": address to the U.S. Senate, January 22, 1917; "open covenants . . .": Fourteen Points, January 8, 1918.

67   "President Wilson is the . . .": quoted in *A Time for Angels*, p. 66; "the whole national . . .": quoted in ibid., p 71; "the president seemed . . ." quoted in ibid., p. 74.

68   "I wish to compare . . .": quoted in ibid., p 73. "left his pedestal . . .": ibid., p. 60; "Mr. Wilson has no . . .": quoted in ibid., p. 61; "one-tenth of its . . .": ibid., p. 126.

69   "in this conference . . .": quoted in ibid., p. 124.

70   "no safeguarding . . . sovereign rights . . . renunciation of the . . .": quoted in ibid., pp. 99–100; "we are invited . . .": quoted in ibid., p. 101.

71   "On the evening . . .": ibid., p. 146; synopsis and quotes from Woodrow Wilson's speech defending the League, from "Appeal for Support of the League of Nations," quoted from *The Public Papers of Woodrow Wilson*, pp. 30–44.

72   "Every lawyer will . . .": ibid.

73   "the United States . . .": quoted in *A Time for Angels*, p. 149.

Al Gore

Primary sources about Al Gore include his book *Earth in the Balance* (Boston: Houghton Mifflin, 1992), the Democratic Party debates between Gore and Bradley, and the presidential debates between Gore and Bush (of which extended excerpts can be found at *http://www.pbs.org/newshour/election2000/speeches/index-preconvention .html*). Gore biographies include Bob Zelnick, *Gore: A Political Life* (Washington: Regnery, 1999); Bill Turque, *Inventing Al Gore* (Boston: Houghton Mifflin, 2000); and David Maraniss and Ellen Nakashima, *The Prince of Tennessee* (New York: Simon and Schuster, 2000).

page

76   "my earliest lessons . . .": *Earth in the Balance*, p. 2; "mother's troubled response . . .": ibid., p. 3.

77   "shocked": ibid., p. 6; "for every 1 percent decrease . . .": ibid., p. 87; "the real danger . . . the whole global . . .": ibid., p. 97; "a global Marshall plan": ibid., p. 295; "cooperative effort": ibid., p. 302; "constraints": ibid., p. 302.

78   "public attitudes are . . . proposals which are . . .": ibid, p. 305; "to try to . . . as a political . . . principal focus": ibid., p. 8; "esoteric": quoted in ibid., p. 8; "in the eyes . . . not even peripheral": quoted in ibid., p. 8; "this problem of . . . Senator Gore just . . .": quoted in *Gore: A Political Life*, p. 157.

79   "the people are ready . . .": quoted in *Inventing Al Gore*, p. 205; "I began to doubt . . .": Gore, *Earth in the Balance*, pp. 8–9; "Mark Twain's cat": Twain's words: "We should be careful to get out of an experience only the wisdom that is in it—and stop there, lest we be like the cat that sits down on a hot stove lid. She will never sit on a hot stove lid again and that is well; but also she will never sit on a cold one anymore." "came to downplay it": *Earth in the Balance*, p. 9; "not a single word . . .": ibid.

80   "I didn't do . . . The harder truth . . .": ibid; "an environmental Paul Revere": quoted in *Inventing Al Gore*, p. 233; "I have become . . . when caution breeds . . .

a good politician . . .": *Earth in the Balance*, p. 15; "muffled . . . were nervous. . . .": *Inventing Al Gore*, p. 257.

81   "he carefully distanced . . .": ibid., p. 257; "ozone man" and "bizarre": quoted in ibid., p. 258.

82   "Al Gore's Silent . . .": ibid., p. 334; "one of the . . .": *The New York Times*, June 22, 1997.

83   "I believe very . . .": first Democratic candidates' debate, Durham, New Hampshire, January 6, 2000; "I think that we . . .": second Democratic candidates' debate, Apollo Theater, New York City, February 22, 2000; "Now that Gore . . .": "Is Al Gore a hero or a traitor?" *Time*, April 26, 1999.

84   "succeeded in driving . . .": *New York Times*, November 3, 2000.

85   "for Mr. Gore to . . .": ibid.

II. TRIANGULATE

George W. Bush

Primary sources include Bush's autobiography, *A Charge to Keep* (New York: HarperCollins, 1999), written with Karen Hughes, as well as the texts of significant public remarks: his speech declaring his candidacy for president (Cedar Rapids, Iowa, June 12, 1999); the Republican primary debates (Phoenix, Arizona, December 6, 1999; Columbia, South Carolina, January 7, 2000; Calvin College, Grand Rapids, Michigan, January 11, 2000; Manchester, New Hampshire, January 26, 2000); and the presidential debates (University of Massachusetts, Boston, Massachusetts, October 3, 2000; Wake Forest University, Winston-Salem, North Carolina, October 12, 2000; Washington University, St. Louis, Missouri, October 17, 2000. Also consulted were Bush's speech at Bob Jones University, Greenville, South Carolina, February 2, 2000, and his letter to New York's Cardinal John J. O'Connor responding to criticism over the Bob Jones visit (February 25, 2000); Bush's speech at the NAACP annual convention, Baltimore, Maryland (July 10, 2000); and Bush's acceptance address at the Republican National Convention (August 3, 2000).

page

97   "A conservative philosophy . . . Some who would . . .": *A Charge to Keep*, p. 236; "prove that someone . . . an old era of . . . politics, after a . . . a fresh start . . .": Cedar Rapids speech, June 12, 1999; "done by churches . . . a quiet river . . .": *A Charge to Keep*, p. 232; "to make a stone of the heart": from "Easter 1916" by William Butler Yeats, *The Poems of William Butler Yeats*, the Variorum edition, Peter Allt and Russell Alspach, eds. (New York: Macmillan, 1957), p. 394; "I know this approach . . . I am proud . . .": Cedar Rapids, June 12, 1999.

98   "I worked to reform . . .": *A Charge to Keep*, p. 236; "match a conservative . . ." Cedar Rapids, June 12, 1999; "everyone belongs . . ." inaugural address, January 20, 2001; "It is conservative . . . It is compassionate . . .": Cedar Rapids, June 12, 1999.

99   "we are now . . .": nomination acceptance speech, Republican National Convention, August 3, 2000; "The seeds of my decision . . .": *A Charge to Keep*, p. 136;

"tolerance and respect . . . a uniter . . . morally grounded . . . a much greater . . . our prosperous . . .": letter to Cardinal O'Connor, February 25, 2000.

100   "shocked" and "horrified" and "directed at America's own . . .": *A Charge to Keep*, pp. 48–9; "the history of the Republican . . .": NAACP annual convention address, July 10, 2000.

101   "Let me start . . . I think that's . . .": Republican candidates' debate, June 12, 1999; "my top priority": NAACP annual convention address, July 10, 2000; "it is conservative . . .": Cedar Rapids, July 12, 1999.

102   "There has to be a consequence . . . When we find . . .": third presidential debate, October 17, 2000.

104   "sought guidance through . . .": *A Charge to Keep*, p. 154; "And what would . . . I support the . . .": Republican candidates' debate, January 10, 2000.

105   "we will reduce . . . The surplus is not . . . America's armed forces . . .": Republican National Convention address, August 3, 2000; "a billion dollar . . .": first presidential debate, October 3, 2000; "to guard against . . .": *A Charge to Keep*, p. 240.

106   "a man of . . .": ibid., p. 177.

Bill Clinton

Primary sources on Clinton include his first acceptance speech at the Democratic National Convention (July 16, 1992), his keynote speech to the Democratic Leadership Council (Cleveland, Ohio, May 6, 1991), and the Democratic candidates' debates (St. Louis, Missouri, October 11, 1992; University of Richmond, Richmond, Virginia, October 15, 1992), as well as Clinton's first and second inaugural addresses (January 20, 1993, and January 20, 1997), and his State of the Union addresses. The author also borrowed from his own book *Behind the Oval Office* (New York: Random House, 1997).

page

113   "Now our new choice . . . Too many of the people . . . the very burdened middle class . . .": Democratic Leadership Council speech, Cleveland, Ohio, May 6, 1991.

114   "often cited to . . ." : *Washington Post*, October 5, 1992, p. A6. "end welfare as . . . you have a responsibility . . .": nomination acceptance speech, Democratic National Convention, July 16, 1992.

115   "New Covenant . . . coddling tyrants": ibid.; "If black people . . . if you took the words . . .": Clinton interview with Jackie Judd, June 17, 1992.

118   "the era of big . . ." : State of the Union address, January 1996.

119   "five fundamental . . . for twelve years . . ." : televised address to the nation, June 13, 1995.

121   "The questions were . . .": *Behind the Oval Office*, p. 212.

122   "As times change . . ." : second inaugural address, January 20, 1997.

123   "unsparing in their . . . playing right into . . . turning Medicare into . . .": *New York Times*, June 14, 1995, pp. A1 and A19.

124   "mend it, but . . .": quoted in George Stephanopoulos, *All Too Human* (Boston: Little, Brown, 1999), p. 374.

François Mitterrand

Secondary sources on Mitterrand include Wayne Northcutt, *Mitterrand: A Political Biography* (New York: Holmes and Meier, 1992); Julius Friend, *The Long Presidency: France in the Mitterrand Years, 1981–1985* (Boulder, Colorado: Westview, 1998); and Vivien Schmidt, *From State to Market? The Transformation of French Business and Government* (New York: Cambridge University Press, 1996), along with Schmidt's article "Business, the state and the end of dirigsme" in *Chirac's Challenge: Liberalization, Europeanization and Malaise in France* (New York: St. Martin's Press, 1996).

page
127 "a biography of . . . Mitterrand that I'd read . . .": in this I refer to *Mitterrand: A Political Biography*.
129 "The nationalizations . . .": Mitterrand press conference, September 18, 1981, quoted in *The Long Presidency*, p. 29; economic statistics on nationalization from *From State to Market?*, pp. 107–8.
130 "a serious drain . . . Within a year . . . fifty-six billion . . . Profit margins . . .": ibid., p. 108; "relatively little . . .": ibid., p. 109; "If they accepted . . .": quoted in Schmidt, *From State to Market?*, p. 111.
132 "fraught with difficult . . . with the day-to-day . . . ambitious, aggressive . . . At the first postelection Council . . .": *Mitterrand*, p. 225; "The transatlantic example . . .": *The Long Presidency*, p. 91.
133 "Publicists described Ronald . . ." ibid., p. 91.
134 "anticipated de-nationalizing . . . if this pace . . .": *From State to Market*, pp. 153–4; "The statesman's duty . . .": *Years of Upheaval* (Boston: Little, Brown, 1982), p. 169.
135 "He hammered hard . . .": *Mitterrand*, p. 254; "The power-sharing arrangement . . .": ibid., p. 272.
136 "cohabitation enabled Mitterrand . . .": ibid., p. 273; "had no laundry list . . .": *The Long Presidency*, p. 111.

Nelson Rockefeller

Rockefeller's own book *Unity, Freedom and Peace: A Blueprint for Tomorrow* (New York: Random House, 1968) is a primary source. Secondary sources included the biography *The Imperial Rockefeller* (New York: Simon and Schuster, 1982) by Joseph E. Persico, Rockefeller's speechwriter, as well as James Poling, *The Rockefeller Record* (New York: Thomas Crowell, 1960), and Theodore S. White, *The Making of the President 1964* (New York: Atheneum, 1965).

page
139 "great-grandfather Spelman . . .": *The Making of the President 1964*, p. 74.
140 "I am proud of . . ." : quoted in Poling, *The Rockefeller Record*, p. 134; "the nation's richest state . . .": quoted in ibid., p. 40; "We cannot speak . . .": quoted in ibid., p. 51.
141 "I can see no . . ." : quoted in the *New York Times*, May 14, 1972, p. A62.
142 "New York, headquarters . . ." : *New York Times*, July 13, 1964, p. A18.

143 "One can no more . . .": *The Making of the President 1964*, p. 74; "disbanded in anger . . . I ain't going to vote . . .": quoted in ibid., p. 80.

144 "deliver a minority . . .": *The Imperial Rockefeller*, p. 65; "outburst . . . boos . . . Some of you . . ." quoted in the *New York Times*, July 15, 1964, p. A1.

145 "Señor Rockefeller . . . I was in the wrong party . . . tried to get me . . . I would rather try . . ." : quoted in *The Imperial Rockefeller*, pp. 80–81.

147 "hawk . . . communism [as] . . .": ibid., p. 72.

148 "There were about . . .": quoted in ibid., p. 258; "ritual act of . . .": ibid., p. 274.

III. DIVIDE AND CONQUER

Abraham Lincoln

Primary sources include the texts of the Lincoln-Douglas debates, published in *The Complete Lincoln-Douglas Debates of 1858*, edited by Paul Angle (Chicago: University of Chicago Press, 1991), a volume especially worth reading for the excellent introduction by Lincoln scholar David Zarefsky. Secondary sources included Zarefsky, *Lincoln, Douglas and Slavery* (Chicago: University of Chicago Press, 1990); Robert W. Johannsen, *Stephen A. Douglas* (Oxford: Oxford University Press, 1973); and William Baringer, *Lincoln's Rise to Power* (Boston: Little Brown, 1937). Other biographies include Carl Sandburg's volumes on Lincoln, particularly *Lincoln: The Prairie Years* (New York: Harcourt, Brace, 1926); Eric Foner's *Free Soil, Free Labor, Free Men*; and Pulitzer Prize winner David Herbert Donald's *Lincoln* (New York: Simon and Schuster, 1995)

page
154 "had little doubt . . .": Donald, *Lincoln*, p. 253.

155 "it will be in despite . . . The Lecompton crisis . . .": quoted in *Stephen A. Douglas*, p. 613.

157 "to be able to . . .": quoted in *Abraham Lincoln: The Prairie Years*, p. 138; "hatred to the institution . . . ultimate extinction . . . as a wrong . . .": quoted in *Free Soil, Free Labor, Free Men*, p. 215; "If here be a man . . .": *The Complete Lincoln-Douglas Debates of 1858*, p. 391.

158 "He has never said . . ." : quoted from *Political Debates Between Lincoln and Douglas*, (1897), p. 461; "each man . . .": ibid., p. 447; "if one man . . .": ibid., p. 491; "This declared indifference . . .": speech in Peoria, 1854, read by Lincoln during the debate with Douglas at Ottawa, August 21, 1858, quoted from *The Complete Lincoln-Douglas Debates of 1858*; "Lincoln was frank . . . Each man . . .": quoted in *Lincoln, Douglas and Slavery*, p. 186.

159 "Can the people . . ." : from the debate with Douglas at Freeport, Illinois on August 27, 1858, quoted from *The Complete Lincoln Douglas Debates of 1858*, p. 143; "I answer emphatically . . .": ibid., p. 152; "Douglas' victory . . ." from foreword to ibid., p. xliv.

160 "raised a storm . . .": *Abraham Lincoln: The Prairie Years*, p. 142; "Lincoln showed his . . .": ibid., pp. 154-5; "I am after larger game . . .": quoted in ibid.

161 "It is the duty . . .": quoted in *Stephen A. Douglas*, p. 788; "Douglas' 'Norfolk Doctrine' was . . . his transition to . . . boldest defiances and . . .": ibid., p. 789; "all questions pertaining . . .": quoted in ibid., p. 736.

162 "If Douglas and . . .": quoted in ibid., p. 744; "portrayed as Lincoln's . . . bristle with armed men . . . Douglas did well . . .": quoted in ibid, p. 791.

Richard Nixon, 1968

Primary sources on Richard Nixon include *RN: The Memoirs of Richard Nixon* (New York: Grosset and Dunlap, 1978) and *Six Crises* (Garden City, New York: Doubleday and Co., 1962). Secondary sources include Stephen Ambrose, *Nixon: The Education of a Politician 1913–1962* (New York: Simon and Schuster, 1988) and *Nixon: The Triumph of a Politician 1962–1972* (New York: Simon and Schuster, 1989); Jonathan Aitken, *Nixon: A Life* (Washington, D. C.: Regnery, 1983); Herbert Parmet, *Richard Nixon and His America* (Boston: Little, Brown, 1990); Theodore S. White, *The Making of the President 1968* (New York: Atheneum, 1969), Earl Mazo with Stephen Hess, *Nixon: A Political Portrait* (New York: Harper and Row, 1968); James Humes, *Nixon's Ten Commandments of Statecraft* (New York: Scribner, 1997), and George H. Mayer, *The Republican Party 1854–1966* (New York: Oxford, 1967).

page
165 "pink right down to . . . if she had her . . .": quoted in Nixon: *The Education of a Politician 1913–1962*, p. 218; "The BIG LIE . . ." quoted in ibid., p. 215; "I used a phrase . . .": *RN*, p. 110; "khaki clad president . . .": ibid., p. 112.
166 "I was as friendly . . .": quoted in *Nixon: A Political Portrait*, p. 296; "Republicans must not go . . .": quoted in ibid., p. 300; "How many more American . . .": quoted in ibid., p. 301.
167 "doesn't serve his country . . . overnight Lyndon Johnson . . .": quoted in ibid., p. 301; "I knew if we let . . .": quoted in Doris Kearns [Goodwin], *Lyndon Johnson and the American Dream* (New York: Harper and Row, 1976), p. 264.
168 "a distorted picture of Asia . . .": quoted in *Nixon: A Political Portrait*, p. 306; "the Nation is in grave . . .": quoted in ibid., pp. 306–7; "Nixon is neither . . .": quoted in *The Making of the President 1968*, p. 127; "brainwashed": *Nixon: A Political Portrait*, p. 306.
169 "We're going to win . . .": address to the Republican National Convention, August 8, 1968; "I shall not . . .": *The Making of the President 1968*, p. 124.
170 "At one point [Humphrey] . . .": *RN*, p. 318.
171 "American disillusionment . . . We simply cannot . . .": quoted in *Nixon: A Political Portrait*, pp. 309–10; "I am proud to . . .": nomination acceptance speech, Republican National Convention, August 8, 1968; "plan to end the war": quoted in *Richard Nixon and His America*, p. 506.
172 "secret plan": ibid., p. 506; "although Nixon's critics . . .": *Nixon's Ten Commandments of Statecraft*, p. 107; "was not secret . . .": quoted in *Richard Nixon and His America*, pp. 506–7.
173 "The confrontation in . . .": *The Making of the President 1968*, p. 345; "a man with his fingers stuck . . .": quoted in *The Making of the President 1968*, p. 335; "felt [Humphrey's] cause was . . . Crossroads of Lamentation": ibid., p. 337.
174 "an acceptable risk for peace": quoted in *RN*, p. 318; "maybe a frail straw . . . Richard the chickenhearted . . . I was determined . . .": quoted in ibid., p. 319; "bring us together": *Richard Nixon and His America*, p. 529; "When did you start to . . .": quoted in *The Making of the President 1968*, p. 360.

175 "he had decided to . . . I thought to myself . . . Johnson was making . . .": *RN,* pp. 322–3.

176 "cables intercepted by . . .": *Richard Nixon and His America,* p. 519; "Nixon was fighting back . . . embarrassing position": ibid., p. 520; "the prospects for peace . . .": quoted in *RN,* p. 329; "thousands of supplies . . .": quoted in *Nixon's Ten Commandments of Statecraft,* p. 107.

Thomas E. Dewey

Primary sources include Truman's memoirs, *Years of Trial and Hope 1946–1952* (New York: Doubleday, 1956), as well as the forty-three-page memo of November 19, 1947, in which Truman adviser Clark Clifford outlined positions that Truman should take on civil rights, labor, the farm vote, etc.. The memo can be found at *http://www.whistlestop.org; shstudy=_collections/1948campaign/large/docs/strategies/cam1-3.htm.* Secondary sources included Irwin Ross, *The Loneliest Campaign* (New York: New American Library, 1968); Richard Norton Smith, *Thomas E. Dewey and His Times* (New York Simon and Schuster, 1982); David McCullough, *Truman* (New York: Simon and Schuster, 1992); Clark Clifford with Richard Holbrooke, *Counsel to the President* (New York: Random House, 1991); Herbert Brownell, *Advising Ike* (Lawrence, Kansas: University Press of Kansas, 1993); and "Thomas Dewey and the Campaigns of 1944 and 1948" in Kenneth Thompson, ed., *Lessons from Defeated Presidential Candidates* (Lanham, Maryland: UPA, 1994).

page
179 "the longest, most . . .": *Truman,* p. 493.
180 "had to borrow a car": ibid., p. 503; "You can't make . . .": quoted in ibid., p. 506.
181 "a red scare was . . . so divided between . . .": quoted in ibid., p. 521; "implored his countrymen . . . to behave like . . .": quoted in *Thomas E. Dewey and His Times,* p. 491; "the tragic fact . . . Communists and fellow . . .": quoted in *Truman,* p. 674; "one of that vast group of . . .": quoted in *Thomas E. Dewey and His Times,* p. 442.
182 "accepted a reporter's . . .": ibid., p. 506; "I am not worried . . .": quoted in *Truman,* p. 552; "baloney . . . political pressures were such . . .": quoted in ibid., p. 553; "Republicans are now taking . . . The Republicans are . . .": quoted in ibid., p. 553.
183 "I'm tired of babying the . . .": quoted in *Thomas E. Dewey and His Times,* p. 450; "Under Soviet direction . . .": *Years of Trial and Hope,* p. 98; "the seeds of totalitarian regimes . . .": quoted in ibid., p. 106.
184 "The tougher we get . . .": quoted in Ross, *The Loneliest Campaign,* p. 15; "The world is hungry and . . .": quoted in ibid., p. 147; "a curious mixture . . .": quoted in ibid., p. 147; "If the Democratic Party . . .": quoted in ibid., p. 148.
185 "At the center . . .": ibid., p. 18; "most of the Cabinet . . .": Clifford quoted in ibid., p. 19; "Henry Wallace will be . . . the independent and progressive . . . The Negro vote . . .": Clifford memo of November 19, 1947.
186 "assiduous and continuous . . . because he controls . . . the northern Negro . . . unless the administration . . . It is inconceivable . . .": ibid; "modern, compre-

hensive . . . anti-lynching legislation . . .": *The Loneliest Campaign*, p. 61; "the most astounding presidential . . .": quoted in Nadine Cohodas, *Strom Thurmond and the Politics of Southern Change* (New York: Simon and Schuster, 1993), p. 188; "The leaders of . . .": *Years of Trial and Hope*, p. 180.

187   "Four days later, on . . .": *The Loneliest Campaign*, p. 63; "Truman is only following the . . .": quoted in *Years of Trial and Hope*, p. 183; "influence . . . to have the . . . the southern states are . . . in the bag . . .": quoted in *The Loneliest Campaign*, p. 64; "I was perfectly willing . . .": *Years of Trial and Hope*, p. 184; "a four party race . . .": *Advising Ike*, p. 80.

188   "The greatest achievement . . .": *Years of Trial and Hope*, pp. 221–22.

189   "By November 1946 . . .": Barton J. Bernstein and Allen J. Matusow, *The Truman Administration, A Documentary History* (New York: Harper and Row, 1966), p. 132; "The Wall Street reactionaries . . .": speech at Dexter, Iowa, September 18, 1948; "you can analyze figures . . .": quoted in *Advising Ike*, p. 84; "except for Wilson . . .": Clifford memo of November 19, 1947.

190   "Wise Men . . . everything in their . . .": *Counsel to the President*, p. 4; "the move to recognize Israel . . .": *Advising Ike*, p. 83; "Republican propaganda . . . The nation is already . . .": Clifford memo of November 19, 1947.

191   "Clark Clifford . . . devised a clever . . .": quoted in *Lessons from Defeated Presidential Candidates*, p. 106.

192   "In retrospect . . .": *Advising Ike*, p. 81; "He sought to . . .": ibid., p. 80; "The rift in the . . . the failures of the . . . basically, Truman turned . . .": ibid, p. 82; "We thought Dewey . . .": quoted in *Lessons from Defeated Presidential Candidates*, p. 105.

193   "There were really . . .": ibid.; "prevented the party . . .": *Advising Ike*, p. 70; "soporific . . . bland": *Counsel to the President*, p. 225. "looked like a devil . . .": quoted in *Lessons from Defeated Presidential Candidates*, p. 109; "He was candid with everyone . . . His wife didn't . . .": quoted in ibid., pp. 109–110; "Instead of the big cheer . . .": quoted in ibid., p. 100.

194   "adopted the classic strategy . . .": George H. Mayer, *The Republican Party 1854–1966* (New York: Oxford University Press, 1967), p. 472; "I do not know about accommodations . . .": quoted in *Thomas E. Dewey and His Times*, p. 508; "the bridegroom on top": though Longworth was credited with popularizing the phrase, it was apparently coined by Ethel Barrymore, according to McCullough in *Truman*, p. 672. "crackpots . . . part of the contemptible . . .": quoted in *Thomas E. Dewey and His Times*, p. 535; "the Communist infiltrated . . .": *The Republican Party 1854–1966*, p. 472; "Communists perform the . . .": quoted in *The Loneliest Campaign*, p. 153; "the communists are the . . .": quoted in ibid., p. 162; "Wallace's argument that . . .": ibid., p. 156.

195   "civil rights program . . .": ibid., p. 156; "The International Ladies Garment . . . lost soul . . . quixotic politics could": ibid., pp. 152–53; "one of the reasons . . .": quoted in Richard Walton, *Henry Wallace, Harry Truman and the Cold War* (New York: Viking Press, 1976), p. 231; "anyone Thurmond attacked . . .": ibid., pp. 230–31; "The Crossley Poll . . .": *The Loneliest Campaign*, p. 240.

IV. REFORM YOUR OWN PARTY

Tony Blair

Primary sources for Tony Blair include his book *The Third Way: New Politics for a New Century* (London: Fabian Society, 1998) and *New Britain* (Boulder, Colorado: Westview Press, 1997), which includes speeches and articles by Blair. Secondary sources included John Rentoul, *Tony Blair* (London: Little, Brown, 1995); Jon Sopel, *Tony Blair: The Moderniser* (London: The Penguin Group, 1995); Peter Mandelson and Roger Liddle, *The Blair Revolution* (London: Faber and Faber Ltd., 1996); and *Labour's Last Chance?: The 1992 Election and Beyond*, by Anthony Heath, Roger Jowell, John Curtice, and Bridget Taylor (Cambridge: Dartmouth Publishing Co, 1994). For background on Blair's Labour Party see Keith Laybourn's *A Century of Labour: A History of the Labour Party 1900–2000* (Great Britain: Sutton Publishing Ltd., 2000), and also David Butler and Dennis Kavanagh's *The British General Election of 1997* (London: Macmillan; New York: St. Martin's Press, 1997).

page
201 "the Labour Party was . . .": *A Century of Labour*, p. 112.
202 "fairness, not favours.": quoted in *Tony Blair: The Moderniser*, p. 109; "By the 1970s . . .": Geoffrey Wheatcroft, "The Paradoxical Case of Tony Blair," *Atlantic Monthly*, June 1996; "Labour's failure to . . .": *A Century of Labour*, p. 116.
203 "lacked a strong . . .": ibid.; "the longest suicide . . .": quoted in ibid., p. 118; "to the middle ground . . .": quoted in ibid., p. 119.
204 "To deal with . . .": ibid., pp. 120–21; "Both the [cities'] . . .": ibid., p. 121; "loony, Labour left . . . Despite all the rhetoric . . .": quoted in ibid., p. 123.
205 "basically a centrist . . .": quoted in *Tony Blair*, p. 82; "the Left's position . . .": ibid., p. 88; "betraying socialist principles . . . The Labor Party needs to . . . position [after 1983 was] . . . the image of . . .": quoted in ibid., p. 146.
206 "profound changes in . . . the key to Mrs. . . .": quoted in ibid., p. 183; "Labour's links with . . .": *Tony Blair: The Moderniser*, p. 106.
207 "there were times . . . an open door policy . . . All the union barons": quoted in ibid., pp. 110–11.
208 "The issue today . . .": quoted in ibid., p. 117; "For the ordinary . . .": ibid., p. 111; "policy clarification.": ibid.
209 Statistics on Labour candidates' attitudes vs. voters' preferences: *Labour's Last Chance?*, pp. 178–83.
210 "Mr. Major did not win . . .": quoted in *A Century of Labour*, p. 128; "Voters just did not . . .": *The Sun*, quoted in ibid., p. 128; "The worry of the electorate . . .": quoted in *Tony Blair: The Moderniser*, p. 138; "I just wish I knew": quoted in ibid., p. 152.
211 "It is now time . . .": quoted in ibid., p. 198
212 "new right had struck . . .": quoted in ibid., pp. 209–10; "a high tax . . . low success . . .": quoted in ibid., p. 211; "sacking [of] incompetent . . .": quoted in ibid., p. 213; "They are not . . .": quoted in ibid., p. 214; "Unions should do the job . . .": quoted in "The Paradoxical Case of Tony Blair," *Atlantic Monthly*, June 1996; "talked in phrases . . .": Tony Blair, p. 9.
213 "only a tiny . . ." *Tony Blair: The Moderniser*, p. 217; "a party has been . . .": "The

Paradoxical Case of Tony Blair," *Atlantic Monthly*, June 1996; "This man, our new . . . destinies of our . . . New Labour . . . I will tell you . . . join us in this . . . through courage . . .": quoted in *Tony Blair*, pp. 4–6.

214 "To secure for . . .": *Tony Blair: The Moderniser*, p. 271; "a competitive market . . . reflex answer to . . . We do not believe . . .": quoted in ibid., pp. 277–8; "In the end . . .": quoted in *A Century of Labour*, p. 138; "showed me what . . .": quoted in Bruce Wallace, "Blair's Blowout," *Maclean's*, May 12, 1997.

215 "A Gallup poll conducted soon after . . .": *Tony Blair: The Moderniser*, pp. 246–49; "two thirds of the electorate . . .": "Striding Clear, British Labor Party Gaining Support," *Economist*, October 1994; "77 percent of the British public . . .": *The British General Election of 1997*, pp. 232–33; "If we can force a draw . . ." quoted in ibid., p. 60.

216 "twenty-two tax . . .": quoted in ibid., p. 60; "bad luck will be . . .": quoted in ibid., p. 23; "The Conservatives have . . .": quoted in ibid., p. 81; "The election will be a . . .": quoted in ibid., p. 84; "I did not spend . . .": quoted in ibid., p. 100; "the most discredited . . . It is 24 hours . . .": quoted in ibid., p. 112; "triumph . . . massacre . . . landslide": quoted in ibid., p. 244.

217 "Voters trusted Labour . . .": ibid., p. 231; "There was no reason left . . .": quoted in ibid., p. 233.

Junichiro Koizumi

Koizumi's rise to power is recent enough that most of the relevant research comes from newspapers and magazine coverage; for my account I have drawn largely on the reporting of the *New York Times*, the *Japan Times*, the *Wall Street Journal*, the *Economist*, and other periodicals. Besides the articles specifically cited below, other helpful pieces from the *Japan Times* included: "Mori advances date of LDP presidential election," March 14, 2001; "Koizumi hints at bid for LDP presidency," March 21, 2001; Ryurichiro Hosokawa, "Lack of leaders is destroying the LDP," March 31, 2001; "Hashimoto, Koizumi eye no. 1 spot," April 7, 2001; "Hashimoto's faction seen bidding for votes via offers of party posts," April 20, 2001; various articles, April 26, 2001; Hugh Cortazzi, "A real chance for change?," April 30, 2001; Toshi Maeda, "LDP candidates hope to ride on Koizumi's wave," July 12, 2001.

page
221 "a secret deal . . .": Kenzo Uchida, "Time running out for Mori," *Japan Times*, February 23, 2001; "country of gods . . .": quoted in Minoru Tada, "Mori's time is running out," *Japan Times*, March 2, 2001; "a serious additional . . .": *Japan Times*, March 4, 2001.

222 "the indications are . . .": Kenzo Nabeshima, "The LDP just doesn't get it," *Japan Times*, March 9, 2001; "The way a prime minister . . .": quoted in "Opposition tells Mori to resign now, not later," *Japan Times*, March 13, 2001; "Now we have to . . . in a way . . .": quoted in "Mori signals intention to resign," *Japan Times*, March 11, 2001; "a group of 40 LDP members . . .": Kanako Takahara, article in *Japan Times*, March 13, 2001.

223 "Wearing white headbands . . . the rebellious group . . .": Kanako Takahara, article in *Japan Times*, March 13, 2001; "the most tragic . . .": quoted in Kanako Takahara, "LDP race: popular appeal vs. vote machines," *Japan Times*, April 19,

2001; "The LDP seems . . . For the LDP . . .": Keizo Nabeshima, "A turning point for the LDP," *Japan Times*, April 11, 2001.

224   "All I can do . . .": quoted in "Mori signals intention to resign," *Japan Times*, March 11, 2001; "I want to explore ways . . .": quoted in "Koizumi May Run," *Japan Times*, March 30, 2001; "He resembles JFK . . .": quoted in Howard W. French, "Japan's New Superstar Redefines Its Politics," *New York Times*, June 24, 2001.

225   "So I'm an eccentric? . . .": quoted in Victoria James, "Triumph of an accidental hero," *New Statesman*, July 9, 2001; "I'd like to see people . . .": quoted in "Politicians are afraid of elections and losing votes," *Time Asia*, November 27, 2000; "addicted to debts . . . the Japanese government borrowed . . .": quoted in "LDP presidential candidates spar over cure for economy," *Japan Times*, April 16, 2001.

226   "hundreds of thousands . . .": "The Politics of Pain," *The Economist*, June 4, 2001; "millions of Japanese . . . shoveled into wasteful . . . of all his . . . without sacred cows . . . resolutely implementing . . .": Brian Bremner, Robert Neff, Ken Belson, Julia Lichtblau, and Irene Kunii, "Japan's Reformer," *Business Week*, May 7, 2001; "the time will come . . .": quoted in "Politicians are afraid of elections and losing votes," *Time Asia*, November 27, 2000.

227   "dedicating flowers to Hitler": quoted in "Koizumi again talks reform," *Japan Times*, July 12, 2001; "Even if I asked them . . . some dissident junior . . .": quoted in Kanako Takahara, "LDP race: popular appeal vs. vote machines," *Japan Times*, April 19, 2001; "he looked like . . .": quoted in *Japan Times*, April 12, 2001.

228   "I will vote for . . .": quoted in Kanako Takahara, "LDP race: popular appeal vs. vote machines," *Japan Times*, April 19, 2001; "tainted by charges . . .": quoted in *Japan Times*, April 19, 2001; "The fact that the . . .": quoted in "Koizumi vows to destroy forces blocking reforms," *Japan Times*, April 24, 2001.

229   "I will appoint . . .": ibid.; "a national salvation": quoted in "Koizumi names new cabinet," *Japan Times*, April 27, 2001; "Gambaro . . .": quoted in Bremner, Neff, Belson, Lichtblau, Kunii, "Japan's reformer," *Business Week*, May 7, 2001.

230   "it seems the earth . . .": quoted in ibid.; "an assault on the . . .": "The voters give Koizumi a chance, will the LDP?," *The Economist*, August 4, 2001.

231   "LDP members, especially . . . The LDP can . . .": quoted in Toshi Maeda and Kanako Takahara, "Poll a vote of faith for Koizumi's untested reforms," *Japan Times*, July 30, 2001; "As long as I . . .": quoted in "Japan's New Leader Speaks His Mind, Ruling Party Will Have to Move Toward Me," *Wall Street Journal*, June 29, 2001.

232   "I will carry out reforms . . .": quoted in Douglas Struck and Kathryn Tolbert, "Koizumi Fever Hits Japanese Campaign," *Washington Post*, July 13, 2001.

George McGovern

Primary sources for George McGovern include his autobiography *Grassroots* (New York: Random House, 1977) and *An American Journey: The Presidential Campaign Speeches of George McGovern* (New York: Random House, 1974). Secondary sources included Theodore S. White, *The Making of the President 1972* (New York: Atheneum, 1973); Robert Sam Anson, *McGovern : A Biography* (New York: Holt, Rinehart, 1972);

Gary Hart, *Right from the Start: A Chronicle of the McGovern Campaign* (New York: Quadrangle/New York Times, 1973); and Herbert Parmet, *The Democrats* (New York: Macmillan, 1976).

page

235 "Humphrey had entered no . . .": Theodore S. White, *The Making of the President 1968*, p. 270; "politician's party . . . From clubhouse to city hall . . .": ibid., p. 65; "The whole world is . . .": ibid., p. 299.

236 "about what is going on . . . With George McGovern . . . Fuck you . . . How hard it . . .": quoted in ibid., p. 210; "Abe and the mayor . . .": *Grassroots*, p. 124; "Back and forth . . .": *The Making of the President 1968*, p. 302.

237 "incomprehensible": ibid, p. 274; "all Democratic voters . . .": *Grassroots*, p. 134.

238 "anyone capable of . . . the real threat . . . was needlessly . . . wasn't happy with . . .": quoted in *McGovern: A Biography*, p. 60; "double-crossed": quoted in Jerry Wurf, "Running without George Meany," *New Republic*, August 5 and 12, 1972, p. 22; "When parties have been . . .": quoted in *McGovern: A Biography*, p. 248; "participatory democracy . . . the party would only . . . never again would . . .": *Grassroots*, p. 139.

239 "the party which nominated . . .": quoted in *Grassroots*, p. 141; "a thankless task . . . getting out of it . . .": quoted in *McGovern: A Biography*, p. 246; "the flak began . . . 'livid' over the notion . . . the unions proceeded . . .": ibid., 249; "an arm of the . . .": quoted in *Grassroots*, p. 140; "we have to be willing . . .": speech at Hunter College, New York, December 9, 1971, quoted from *An American Journey*, p. 173.

240 " 'if you're asking for amnesty . . .' . . . rebuffed the reformers . . . it would have been better. . . .": quoted in *Grassroots*, p. 141; "was so infuriated . . . the commission offended some . . . the party's shortcomings . . .": "Democrats: Reform or Die," *Time*, June 27, 1969, p. 18.

241 "noticing how much . . .": *Grassroots*, p. 231.

243 "The Commission on Democratic . . .": speech at Hunter College, New York, December 9, 1971, quoted from *An American Journey*, p. 174; "we will not be weak-kneed . . .": quoted in *McGovern*, p. 248; "We will not be helped . . .": speech at Sioux Falls, South Dakota, January 18, 1971, quoted from *An American Journey*, p. 5; "to the hilt" and "continued to rely . . .": *The Democrats*, p. 301.

244 "In a democratic nation . . . The destiny of . . .": "Come Home, America," nomination acceptance speech, Democratic National Convention, Miami, Florida, July 13, 1972, quoted from *An American Journey*, p. 20.

245 "one thousand percent": quoted in *RN*, p. 663; "I wouldn't have hesitated . . .": quoted in *New York Times*, July 26, 1972, p. 20; "I don't like to think . . . would make me most uncomfortable": quoted in *Grassroots*, p. 215; "It is painfully evident . . .": *New York Times*, July 28, 1972; "the . . . matter ended . . .": *Grassroots*, p. 215; "one rock in the landslide . . . Perhaps that is true . . .": ibid., pp. 216–17; "very, very shabby treatment . . . this fall's election . . .": quoted in ibid., p. 216; "wouldn't touch . . .": "Can McGovern be stopped?," *U.S. News & World Report*, June 19, 1972, p. 14.

246 "he don't stick to his people": quoted in *RN*, p. 672. "for the large block . . .": quoted in "Can McGovern be stopped?," *U.S. News & World Report*, June 19, 1972, p. 16; "completely unacceptable to the . . .": quoted in "Democratic Gov-

ernor's Price for Getting on McGovern Bandwagon," ibid., June 19, 1972, p. 16; "in a presidential campaign . . .": quoted in *RN*, p. 673; "patch up matters with the traditional . . .": *The Democrats*, p. 303

V. USE A NEW TECHNOLOGY

Franklin D. Roosevelt

An important resource throughout this fifth section of the book was the Museum of Television and Radio. with locations in Los Angeles and New York. Its web site is *www.mtr.org/welcome.htm*. A number of good collections of FDR's fireside chat speeches have been published; among them are *Nothing to Fear: The Selected Addresses of Franklin Delano Roosevelt 1932–1945*, edited by B. D. Zevin (Boston: Houghton Mifflin 1946), and *The Roosevelt Reader*, edited with an introduction by Basil Rauch (New York: Rinehart and Co., Inc., 1957); another source is *http://www.mhrcc.org/fdr/ fdr.html*, which reproduces texts as supplied by the FDR Presidential Library.

Exceptional secondary sources on which I have relied include Robert Brown, *Manipulating the Ether: The Power of Broadcast Radio in Thirties America* (Jefferson, N.C. : McFarland & Co., 1998), the best book on the subject; Betty Houchin, *FDR and the New Media* (Chicago: University of Illinois Press, 1990); Donald G. Godfrey and Frederic Leigh, *Historical Dictionary of American Radio* (Westport, Connecticut: Greenwood Press, 1998); and David Kennedy, *Freedom from Fear: The American People in Depression and War, 1929–1945* (New York: Oxford University Press, 1999).

page

254   "countless citizens found . . .": *Manipulating the Ether*, p. 3; "nothing since the . . .": quoted in ibid., p. 9.

255   "Amid many developments of . . .": quoted in ibid., p. 10; "During each [legislative] session . . .": quoted in ibid., p. 28. "Time after time . . .": quoted in ibid., p. 29; "radio's full and unqualified . . ." quoted in ibid., p. 14.

256   "unexpectedly found itself . . .": ibid., p. 15; "the authority of his voice": quoted in ibid., p. 12; "Radio has given a new meaning . . .": *New York Times*, March 13, 1933, quoted in ibid., p. 25; "The traditional practice . . . value of drama . . . dismayed, disheartened . . .": quoted in ibid., p. 30; "I pledge myself . . .": nomination acceptance speech, Democratic National Convention, July 2, 1932; "the greatest debate . . .": quoted in *Manipulating the Ether*, p. 31.

257   "Roosevelt's method was . . .": quoted in ibid., p. 31; "the only thing we have to fear . . .": first inaugural address, March 4, 1933.

258   "this nation will . . .": first inaugural address, March 4, 1933; "the human voice . . .": *New York Times*, June 18, 1933, p. 2, quoted in *Manipulating the Ether*, p. 62; "I do not promise . . . After all . . . there is an element . . .": Fireside Chat on banking, March 12, 1933.

259   "In a voice . . . On Monday the . . .": *Freedom from Fear*, pp. 136–37; "It was as if a wise . . .": quoted in *Manipulating the Ether*, p. 65. "His use of this new . . .": quoted in *FDR and the New Media*, p. 104; "As Brown reports, one study . . .": *Manipulating the Ether*, p. 19.

260   "where's that glass . . . my friends . . .": quoted in *Manipulating the Ether*, p. 18;

"While most radio orators . . .": ibid., p. 19; "were more sensitive . . ." quoted in ibid., p. 20; "My husband had . . .": quoted in *FDR and the New Media*, p. 105; " 'fireside chat' . . . the president likes to think . . .": quoted in *Manipulating the Ether*, p. 19; "Standing on either side . . . picture in his . . .": quoted in ibid.

261   "His magnetic voice has . . .": *New York Times*, June 1933, quoted in ibid., p. 20; "one thing I dread . . . I ought not to appear oftener . . .": quoted in *FDR and the New Media*, p. 105; "one man revolution . . .": *Broadcasting*, quoted in *Manipulating the Ether*, p. 19; "My wife and I . . .": quoted in ibid., p. 60.

262   "It is wholly wrong . . .": Fireside Chat of May 7, 1933; "Are you better off . . .": Fireside Chat of June 28, 1934.

263   "It was your nationwide . . .": quoted in *Manipulating the Ether*, p. 37; "need not have been . . .": quoted in ibid., p. 76.

264   "Until 4:30 this morning . . .": September 3, 1939, quoted in *The Roosevelt Reader*, p. 222.

265   "Before the broadcast, *Fortune* said . . .": quoted in *Manipulating the Ether*, p. 100; "As Roosevelt spoke, attendance at New York City . . .": ibid., p. 94; "There are many among us . . .": Fireside Chat, May 26, 1940; "peacetime army": Fireside Chat, October 29, 1940.

266   "My conscience will . . . if his promise . . .": quoted in *Manipulating the Ether*, pp. 42–43; "I have said this before . . .": radio address, October 30, 1940; "His election for a third . . .": quoted in *Manipulating the Ether*, p. 45; "You would not haggle . . .": press conference, December 17, 1940.

267   "as complicated and . . .": quoted in *Manipulating the Ether*, p. 90; "Never before since Jamestown . . .": quoted from *Nothing to Fear*, p. 247; "initiated two unwarranted . . .": Fireside Chat on September 11, 1941, quoted from ibid., p. 287.

John F. Kennedy/Richard Nixon, 1960

The 1960 Nixon-Kennedy debates were reviewed at the Museum of Television and Radio. The Nixon Presidential Library in Yorba Linda, California, is also a first-rate resource for documents on Nixon's use of the media; the museum has a television that plays the Checkers speech around the clock. Books consulted include Richard Nixon, *RN: The Memoirs of Richard Nixon* (New York: Grosset and Dunlap, 1978) and *Six Crises* (Garden City, New York: Doubleday and Co, 1962); Theodore White, *The Making of the President 1960* (New York: Atheneum, 1961); Stephen Ambrose, *Nixon: The Education of a Politician 1913–1962* and *Nixon: The Triumph of a Politician 1962–1972* (New York: Simon and Schuster, 1987, 1989); Herbert Parmet, *Richard Nixon and His America* (Boston: Little, Brown, 1990); and Jonathan Aitken, *Nixon* (Washington: Regnery, 1983). Also helpful were Gene Wyckoff, *The Image Candidates* (New York: Macmillan, 1968); Marshall McLuhan, *Understanding Media* (New York: McGraw Hill, 1964); Christopher Matthews, *Kennedy and Nixon: The Rivalry that Shaped Postwar America* (New York: Simon and Schuster, 1996); Arthur Schlesinger, Jr., *A Thousand Days* (Boston: Houghton Mifflin, 1965); Theodore Sorensen, *Kennedy* (New York: Harper and Row, 1965); Herbert Parmet, *JFK* (New York: The Dial Press, 1983); Larry Sabato's *The Rise of Political Consultants* (New York: Basic Books, 1981); and Arthur Schlesinger, Jr.'s *Kennedy or Nixon?* (New York: Macmillan, 1960), published during the campaign.

page
270 "willingly be[came] a media": ibid., p. 14; "malaria-ridden . . .": quoted in *Kennedy and Nixon*, p. 133; "visits to Palm Beach . . .": *Kennedy*, pp. 38–9. "the great risk . . .": quoted in ibid., p. 41; "transformed . . .": *Kennedy and Nixon*, p. 139; "enabled Kennedy to . . .": David Burner, *John F. Kennedy and a New Generation* (Boston: Little, Brown, 1988), p. 116.

271 "movie star": quoted in *John F. Kennedy and a New Generation*, p. 34; "what better narrator . . .": quoted in ibid. p. 33.

272 "The Kennedy campaign hired . . .": Larry Sabato, *The Rise of Political Consultants* (New York: Basic Books, 1981), p. 117; "the true Narcissus . . .": McLuhan, *Understanding Media* (New York: McGraw Hill, 1964), p. 11; "used the medium . . .": ibid., p. 336.

274 "The T.V. experts . . . was carefully coached . . .": Edwin Diamond and Stephen Bates, *The Spot: The Rise of Political Advertising on Television* (Cambridge, Massachusetts: MIT Press, 1984), p. 71; "fifty-eight million people . . .": *New York Times*, September 24, 1952, and *Nixon: The Education of a Politician 1913–1962*, p. 289; "Pat doesn't have . . .", "[A] Texas supporter . . .": from Nixon's Checkers speech; "that surge of confidence . . .": *RN*, p. 104; "when I saw the elevator . . .": quoted in *Nixon: The Education of a Politician 1913–1962*, p. 289; "a figure from . . .": *Richard Nixon and His America*, p. 246; "the telephones are . . .": quoted in ibid., p. 249.

275 "hand outstretched . . . you're my boy": *RN*, p. 106; "it isn't what the facts are . . .": quoted in *Nixon: The Education of a Politician 1913–1962*, p. 294; "don't know anything . . .": quoted in ibid., p. 523; "You do all the talking . . .": *New York Times*, July 25, 1959.

276 "[Nixon] had no reason to help . . .": quoted in Earl Mazo and Stephen Hess, *Nixon: A Political Portrait* (New York: Harper and Row, 1968), p. 234; "was convinced he could win . . .": ibid., pp. 234–35; "style was a simple . . . On television, the deep eye wells . . .": *The Making of the President 1960*, p. 277; "an expert in lighting . . .": *Nixon: The Education of a Politician 1913–1962*, p. 559.

277 "white and pasty": ibid., p. 773.

278 "deep scar . . .": *Six Crises*, p. 128; "showmanship" and "statesmanship": *RN*, p. 221; "It is a devastating commentary . . .": *RN*, p. 219; "I had concentrated . . .": *Six Crises*, p. 240; "heard Jack Kennedy's . . .": *Kennedy and Nixon*, p. 149.

279 "for the first time . . .": quoted in *The Spot*, p. 112. "did not want . . .": *The Image Candidates*, p. 21; "relegat[ing] such television . . .": ibid., p. 44; "the torch has been . . .": John F. Kennedy, inaugural address, January 20, 1960.

Lyndon B. Johnson

Sources included Theodore S. White, *The Making of the President 1964* (New York: Atheneum, 1965); Edwin Diamond and Stephen Bates, *The Spot: The Rise of Political Advertising on Television* (Cambridge, Massachusetts: MIT Press, 1984); Tony Schwartz, *The Responsive Chord* (Garden City, New York: Anchor/Doubleday, 1973); Kathleen Hall Jamieson, *Dirty Politics* (New York: Oxford University Press, 1992) and *Packaging the Presidency* (New York: Oxford, 1996); Larry Sabato, *The Rise of Political*

Consultants (New York: Basic Books, 1981); and Stephen Ansolabehere and Shanto Iyengar, *Going Negative* (New York: The Free Press, 1995). Biographies include Robert Dallek's two-volume work, *Lone Star Rising: Lyndon Johnson and His Times 1908–1960* (New York: Oxford University Press, 1991) and *Flawed Giant: Lyndon Johnson and His Times 1961–1973* (New York: Oxford University Press, 1998), as well as Doris Kearns (Goodwin), *Lyndon Johnson and the American Dream* (New York: Harper and Row, 1976).

page

281   "I don't think it would . . .": quoted in *The Spot*, p. 41; "If you give me . . .": quoted in Stephen Ambrose, *Nixon: The Education of a Politician 1913–1962* (New York: Simon and Schuster, 1987), p. 559.

283   "the temporary spokesman . . . out of tune . . . Most Democrats and most . . .": address to the Democratic National Convention, August 27, 1964; "small conventional nuclear . . .": quoted in "Conventional nuclear weapons for NATO urged by Goldwater," *New York Times*, August 26, 1964, p. A29; "western civilization can . . .": address to the Democratic National Convention, August 27, 1964; "stridency and unrestrained . . .": quoted in *New York Times*, August 28, 1964.

284   "I seek the support . . .": quoted in *Packaging the Presidency*, p. 185; "After the Republican . . .": quoted in ibid., p. 186; "If we are not careful . . ." quoted in *Flawed Giant*, p. 174; "Television ads . . .": quoted in ibid., p. 175.

285   "Would you work . . .": quoted in *The Spot*, p. 127; "one, two, three. . . . ten, nine, eight . . .": *The Spot*, p. 128; "These are the stakes . . .": ibid., p. 129; "The commercial evoked . . .": *The Responsive Chord*, p. 93; "Peace little girl. . . .": ibid.; "all three networks. . . . .": Lee Edwards, *Goldwater: The Man Who Made a Revolution* (Washington, D.C.: Regnery, 1995), pp. 300–301.

286   "You got your point across": quoted in *Flawed Giant*, p. 175; "You sure we ought to run it . . .": quoted in ibid., p. 176; "Goldwater should have said . . .": interview with Tony Schwartz, January 24, 2002; "Every time I saw . . .": quoted in *The Spot*, p. 132; "When people hear . . ." ibid., p. 133; "Yeah, but we can't . . .": quoted in ibid; "to put non-paid media in context": *The Responsive Chord*, p. 98; "experiences with television . . .": ibid., p. 144; "My wife loves. . . .": ibid., pp. 44–5.

287   "Do you know . . .": *The Spot*, p. 133; "On at least seven occasions . . .": ibid., p. 137; "this particular phone . . .": ibid., p. 135; "merely another weapon . . .": ibid.

288   "In Your Heart . . .": ibid., p. 140; "reckless": *Flawed Giant*, p. 177; "the breadth of . . .": *Going Negative*, p. 90; "My candidate had been . . .": quoted in *The Spot*, p. 140.

289   "holocaust . . . push the button . . . atomic weapons . . .": quoted in *The Making of the President 1964*, p. 342; "tommyrot": quoted in *The Spot*, p. 140; "What has happened to America?": *The Spot*, p. 146; "so indecisive that . . .": ibid., p. 141; "tell the viewer anything . . .": *The Responsive Chord*, p. 93.

VI. MOBILIZING THE NATION IN TIMES OF CRISIS

Winston S. Churchill, Franklin D. Roosevelt (World War II)

Lyndon B. Johnson (Vietnam)

George W. Bush (The War on Terror)

Sources consulted include: Doris Kearns Goodwin, *Lyndon Johnson and the American Dream* (New York: St. Martin's Press, 1991); Irwin and Debi Unger, *LBJ* (New York: John Wiley and Sons, 1999); Paul Conkin, *Big Daddy from the Pedernales* (Boston: Twayne Publishers, 1986); Robert Dallek, *Flawed Giant: Lyndon Johnson and His Times 1961–1973* (New York: Oxford University Press, 1998); Joseph Califano, Jr., *The Triumph and Tragedy of Lyndon Johnson* (New York: Simon and Schuster, 1991); and David Halberstam, *The Best and the Brightest* (New York: Random House, 1969). The material on Winston Churchill was drawn in large part from the same body of resources as listed in Part I above; for the material on FDR, besides the aforementioned sources, books also include Jeffrey Hacker, *Franklin D. Roosevelt* (New York: Franklin Watts, 1983); Kenneth S. Davis, FDR: *The War President 1940–1943* (New York: Random House, 2000), and, of course, the texts of numerous FDR speeches and radio addresses.

Reviewed with hindsight, George Bush's speeches at the dawn of the War on Terror may someday carry comparable weight. Paramount among them are his remarks of September 12, 2001; the speech "Freedom and Fear Are at War" (September 20, 2001), in which he describes the challenge by calling the terrorists "the heirs of all the murderous ideologies of the 20th century"; "Duty and Sacrifice," October 7, 2001; a speech to the March of Dimes on October 12, 2001; and Bush's October 11, 2001, press conference.

page
297   "I do not at . . .": "Their Finest Hour," speech to House of Commons, June 18, 1940, quoted from *Blood, Toil, Tears and Sweat: The Speeches of Winston Churchill*, edited by David Cannadine (Boston: Houghton Mifflin, 1989), p. 174.
298   "If the British . . .": "Their Finest Hour": June 18, 1940, quoted from *Blood, Toil, Tears and Sweat*, p. 178; "a date which will live . . . long will we . . .": war message to Congress, December 8, 1941, quoted from *The Roosevelt Reader*, p. 300; "Every single man . . .": Fireside Chat, December 9, 1941, quoted from *Nothing to Fear*, p. 306; "the task that we Americans . . .": Fireside Chat, February 23, 1942, quoted from *The Roosevelt Reader*, pp. 308–9; "This war is . . .": Fireside Chat, February 23, 1942, quoted from ibid., p. 304; "in such imperceptible . . .": *The Best and the Brightest*, p. 593.
299   "Seated on the . . .": *Lyndon Johnson and the American Dream*, p. 273; "it would be . . .": *The Best and the Brightest*, p. 456; "even to himself . . .": *Big Daddy from the Pedernales*, p. 262; "without major costs . . . did almost nothing to . . .": ibid., p. 265; "on the road ahead . . . I was about . . .": Fireside Chat, December 9, 1941.
300   "If there is one . . .": quoted in *Lyndon Johnson and the American Dream*, p. 286; "we are not . . .": speech at Akron, October 21, 1964, quoted in *The Triumph*

and Tragedy of Lyndon Johnson, p. 172; "those people out there . . .": quoted in *The Best and the Brightest*, p. 594; "How could Johnson . . .": *Lyndon Johnson and the American Dream*, p. 296; "Johnson's concept of . . . the public . . .": ibid., p. 297; "to avoid clarifying . . .": *The Best and the Brightest*, p. 584.

301 "went into one . . . the most vehement . . .": ibid., p. 587; "filling out a . . .": ibid., p. 599 (the reporter was Douglas Kiker of the *New York Herald Tribune*); "does not imply . . .": quoted in ibid., p. 601; "In effect, the Administration . . .": ibid., p. 609; "I have said . . .": quoted in *The Triumph and Tragedy of Lyndon Johnson*, p. 172; "quarantine": "Quarantine the Aggressor," speech of October 5, 1937; "neutral in thought . . .": Fireside Chat, September 3, 1939.

302 "my garden hose": press conference outlining premises of Lend-Lease program, December 17, 1940; "the first battle . . . I think the . . . that after September . . .": Bush press conference, October 11, 2001; "this is an unconventional war . . . It's not the kind . . .": ibid.

303 "guided by a greater . . . some important things . . . reminded of the . . .": speech to March of Dimes, October 12, 2001; "British nation is . . .": "Westward, Look, the Land Is Bright," BBC speech, April 27, 1941, from *Blood, Toil, Tears and Sweat*, p. 215; "this generation of . . .": Fireside Chat, February 23, 1941; "This nation is mighty . . .": State of the Union address, January 12, 1966; "$5 billion tax increase": *The Triumph and Tragedy of Lyndon Johnson*, p. 112.

304 "the business of war . . . in the end . . .": *Lyndon Johnson and the American Dream*, p. 298; "most people associate . . .": Louis Heren, *No Hail, No Farewell* (New York: Harper and Row, 1970), quoted in *LBJ*, p. 389; "There are men . . .": State of the Union address, January 12, 1966; "casualty lists of . . .": Fireside Chat, December 9, 1941, quoted from *Nothing to Fear*, p. 309; "There will be . . .": ibid.

305 "many other creature . . .": Fireside Chat, April 28, 1942, quoted from ibid.; "the price for . . .": ibid; "Never in the . . .": ibid; "Washington's conduct in . . .": Washington's Birthday Fireside Chat, February 23, 1942, quoted from *The Roosevelt Reader*, p. 303; "The WPA. . . .": *Franklin D. Roosevelt*, p. 87.

306 "the attack yesterday . . .": war message to Congress, December 8, 1941, from *The Roosevelt Reader*, p. 300; "We must share . . .": Fireside Chat, December 9, 1941, quoted from *Nothing to Fear*, p. 306; "Some people may be . . .": "A Long, Hard War," address to joint session of Congress, December 26, 1941, quoted from *Blood, Toil, Tears and Sweat*, p. 229.

307 "Public pronouncements soon . . .": *Big Daddy from the Pedernales*, pp. 252–53; "The enemy is . . . Time is no longer . . .": State of the Union address, January 12, 1966; "By late 1966 . . . The president is . . .": quoted in *The Triumph and Tragedy of Lyndon Johnson*, p. 172; "Why, Lyndon, you . . .": quoted in *The Best and the Brightest*, p. 442; "seemed amusing early . . .": *The Best and the Brightest*, p. 449–50; "might go public . . .": quoted in ibid, p. 552.

308 "No, he won't . . .": quoted in ibid; "I deprecate premature . . .": speech on the war situation, September 30, 1941, quoted from *The Unrelenting Struggle*, compiled by Charles Eade (Boston: Little, Brown, 1942), pp. 265–66; "wars are not won by evacuations": speech to House of Commons, June 4, 1940, from *Blood, Toil, Tears, and Sweat*, p. 292; "We must be . . .": "War Production," speech to House of Commons, July 29, 1941, quoted from *The Unrelenting Struggle*, p. 221; "the mood of Britain . . .": speech on the war situation to House of Commons, September 9, 1941, from ibid., p. 261; "I wish I could . . .": "Forward, Till the

Whole Task Is Done," BBC speech, May 13, 1945, from *Blood, Toil, Tears and Sweat*, p. 265; "The House will remember . . .": speech to House of Commons, September 30, 1941, from *The Unrelenting Struggle*, p. 264.

309   "If I were to . . .": quoted from *The Unrelenting Struggle*, p. 267; "Mr. Bush tried to . . .": editorial, *New York Times*, October 12, 2001; "covert operations secret . . .": address to a joint session of Congress and the American people, September 20, 2001; "did not fully . . .": *Big Daddy from the Pedernales*, pp. 257–58.

310   "sent a message . . .": Martin Gilbert, *Winston S. Churchill: Finest Hour, Volume VI 1939–1941* (London: Heinemann, 1983), p. 315; "succumb to the . . .": ibid., p. 317.

311   "Congress was solidly . . .": *Franklin D. Roosevelt*, p. 79; "84 percent . . ." ibid., p. 87. "Wars are supposed to . . .": quoted in *The Best and the Brightest*, p. 640; "It was very hard . . .": ibid., p. 623.

312   "premium put on . . .": quoted in ibid., p. 595; "If the military and . . .": ibid., pp. 595–96; "Our outlook at . . .": Winston S. Churchill, *Their Finest Hour* (Boston: Houghton Mifflin, 1949), pp. 350–51; "There is no doubt . . .": quoted in *Winston S. Churchill: Finest Hour, Volume VI 1939–1941*, pp. 567–68; "full of the most . . .": quoted in ibid., p. 675.

Index

About the Author

Dick Morris is the author of four previous books, including the *New York Times* bestseller *Behind the Oval Office*. A former campaign advisor to politicians as diverse as President Bill Clinton and Senator Trent Lott, he is currently a political analyst for the Fox News Channel and a columnist for the *New York Post*. Morris divides his time between Connecticut and New York.